Sustainable Tourism, Part B

Editorial Advisory Board

Sustainable Tourism, Part B: A Comprehensive Multidimensional Perspective

EDITED BY

DISHA SHARMA
Amity University Chhattisgarh, India

HAMID ABDULLAH
Atal Bihari Vajpayee Vishwavidyalaya, India

AND

PRAMENDRA SINGH
Lovely Professional University, India

emerald
PUBLISHING

United Kingdom – North America – Japan – India – Malaysia – China

Emerald Publishing Limited
Emerald Publishing, Floor 5, Northspring, 21-23 Wellington Street, Leeds LS1 4DL

First edition 2024

Reprints and permissions service
Contact: www.copyright.com

British Library Cataloguing in Publication Data
A catalogue record for this book is available from the British Library

ISBN: 978-1-83608-327-6 (Print)
ISBN: 978-1-83608-326-9 (Online)
ISBN: 978-1-83608-328-3 (Epub)

Printed and bound by CPI Group (UK) Ltd, Croydon, CR0 4YY

INVESTOR IN PEOPLE

Contents

About the Editors

Dr Disha Sharma is part of the Finance Faculty, currently working as an Assistant Professor at Amity Business School, Amity University Chhattisgarh. She is a qualified UGC-NET & SET, did Master's in three subjects and completed her PhD from Pt. Ravi Shankar Shukla University. She has 6 years of teaching experience. She has published 10 Scopus and four peer-reviewed papers. She has also written seven book chapters in international and Scopus-indexed books. She has successfully edited three books under reputed journals and is currently editing two books on sustainability in tourism with Emerald Publishing. She also has experience as a Co-Principal Investigator in ICSSR and UNICEF Project and is currently working on another project as the Principal Director. She has presented nine research papers at national and international conferences. She is a lifetime member of the Indian Commerce Association and Indian Accounting Association.

Dr Hamid Abdullah, an accomplished academician with over 15 years of experience in academia and industry, currently serves as an Assistant Professor and the Head of the Department of Hotel Management and Hospitality at Atal Bihari Vajpayee University Bilaspur (formerly Bilaspur University) in Bilaspur, Chhattisgarh, India. He holds a PhD from the Institute of Tourism and Hotel Management (ITHM), Bundelkhand University Jhansi, UP India, an MBA specialising in HR and Marketing from the Business School, University of Jammu, and an MTM from Indira Gandhi National Open University. Additionally, he holds a BSc in Hospitality and Hotel Administration from the Institute of Hotel Management (IHM) Srinagar. Dr Abdullah is a qualified UGC NET JRF in Tourism Administration and Management and actively contributes to academia through research papers published in national and international journals, as well as book chapters in edited books. He has also participated in various national and international conferences.

Dr Pramendra Singh completed PhD in tourism and travel management and is currently working as an Assistant Professor in School of Hotel Management and Tourism, Lovely Professional University, India. He has more than 10 years of teaching and research experience. His research area includes heritage tourism, sustainable tourism, tourism and technology and tourism education. He has published more than 25 research papers and book chapters with reputed international and Indian publishers. He is currently editing two books on sustainability in tourism with Emerald Publishing and CABI. He has also been a key resource person for many academic conferences and guest lectures.

About the Editors

About the Contributors

Arun Bhatia received his PhD degree in Travel and Tourism from Central University of Himachal Pradesh, India, in the year 2020. He is currently working as an Assistant Professor in the Department of Travel and Tourism. His research interest area is ICT and Tourism and Gender discourse in Travel and Tourism. He has published many research articles in marketing management in various national and international journals and presented research papers in national and international conferences.

Vaibhav Bhatt is an Assistant Professor in the Department of Tourism and Hospitality Management, Central University of Tamil Nadu, Tiruvarur, Tamil Nadu, India. His research work focuses on formulating sustainable approaches to enhance the tourism sector and aims to bridge the divergence between theoretical concepts and real-world implementation. He has more than 10 years of industry, research and teaching experience and has published over 10 research articles in journals of international/national repute. He is a reviewer of many national/international journals and has also delivered lectures in academic conferences and faculty development programmes.

Kingshuk Biswas is currently employed with the Gujarat State Government, and he looks after renewable energy projects in the State. Earlier, he had an extensive two-decade background with the State Tourism Board, and thus, his passion for tourism remains. Having postgraduate degrees in Economics and Management, he is upbeat about gaining knowledge and experience in the renewable energy sector. Sustainable tourism, policymaking and destination marketing are his key areas of interest.

Farah S. Choudhary is currently working as an Assistant Professor, The Business School, University of Jammu since 2011. PhD in Digital Marketing. Having teaching and research experience of more than 10 years. Area of Interest includes management, tourism, retailing, behaviour finance, portfolio management and digital marketing. Papers published with internationally reputed publishers like Taylor & Francis, IGI Global, Edward Elgar Publishing, ABDC listed, peer-reviewed journals. Attended international/national conferences across the country.

Tsering Chusket is pursuing Master of Travel and Tourism Management (third semester) at Department of Travel and Tourism Management, University of

Ladakh, India. She has been an active member of high-altitude sports organiser for 5 years. Her area of research interest is event planning, event performance measurement and event operation.

Niranjan Deo Pathak is currently an Assistant Professor at Amity Business School, Amity University, Raipur, Chhattisgarh. He is pursuing his PhD (pre-submission completed) from Central University of Jharkhand, Ranchi, Jharkhand. He has completed postgraduation from Vellore Institute of Technology and graduation from Garden City College, Bangalore University. He has authored three research papers in national and international journals and attended four national and international conferences. His area of interest is in tourism, generational study and consumer decision-making.

Tarisai Gundiwa is a tourism and hospitality management principal lecturer in the Department of Tourism and Hospitality Management, Harare Polytechnic College in Zimbabwe. Gundiwa is a holder of BSc and MSc degrees in Tourism and Hospitality Management with University of Zimbabwe. His current research interests focus on sustainable township tourism development, rural tourism development and stakeholder collaboration in sustainable tourism development. He is a member of Tourism Researchers Association of Zimbabwe (TERAZ). Gundiwa worked extensively with Higher Education Examination Council, Zimbabwe, as a tourism and hospitality module developer and a chief examiner.

Maulita Sari Hani specialises in shark and ray conservation and tourism in Indonesia. Her primary focus is on advancing sustainable and inclusive tourism. Her efforts merge scientific expertise, policy implementation and collaborative engagement with diverse stakeholders, strongly emphasising community-based management strategies. Her objective is to protect wildlife, elevating its importance for conservation while simultaneously making it a key tourist attraction, all aimed at enhancing the resilience of local livelihoods. Her research and publications are mainly on manta rays and whale shark tourism.

Anushka Lydia Issac has a background in Engineering [Hons.] and an MBA. Anushka Lydia Issac brings a unique perspective. Holding prestigious credentials such as an FHEA and a Lead IQA Certificate, she possesses the skills and knowledge to design, deliver and assess high-quality learning experiences in higher education. Her passion for understanding people and their personalities has propelled her to excel in communications and public speaking at an international level. As an Associate Faculty and Programme Leader at Westford University College, she actively contributes to the development of students. Her PhD research from the University of Plymouth, UK, focuses on flexible work arrangement.

Aruditya Jasrotia is working as an Assistant Professor in Amity Institute of Travel and Tourism, Amity University Uttar Pradesh. He completed his PhD and MPhil from Central University of Jammu. He has published 13 Scopus publications, 2 patents and 23 publications in total including research papers and book chapters. His broad research areas of interest are smart tourism, tourism marketing, sustainable tourism, tourism policy and planning, tourism products and services. He is also associated as a reviewer for various Scopus-Indexed and ABDC-listed international journals like *International Journal of Tourism Cities*, *Journal of Hospitality, Leisure, Sport & Tourism Education*, *International Journal of Tourism Policy* and *Kybernetes*. Associate Editor of *ESTEEM Journal of Social Sciences and Humanities*.

Shiv Swaroop Jha is currently serving as the Head of Department and Programme Coordinator in Amity College of Commerce, Amity University, Haryana. He has completed his MBA in Financial Management and International Business from Mahamaya Technical University, Noida (2012), and is going to complete his PhD in Management from Bhim Rao Ambedkar Bihar University, Muzaffarpur, Bihar, in 2023. He is the alumni of one of India's most reputed universities, Banaras Hindu University (BHU) Varanasi and Institute of Chartered Accountant of India (ICAI). He joined academia in the year 2012 as a Lecturer in Amity Business School (ABS), Amity University, Haryana. In 2013, he established the Amity College of Commerce. Before this, he has served many reputed consultancy and audit firms as an Auditor and Audit Manager in Kanpur and Delhi, NCR region. He has 20+ publications (peer-reviewed papers and book chapters/ articles) to his credit. He has very rich experience in international taxation, auditing, accounting and financial management. He has performed various statutory, tax, bank and internal audit in the prestigious organisations like Reserve Bank of India, Punjab National Bank, National Highway Authority of India, Indian Oil Corporation Ltd, Dharampal Satyapal Group, Asahi India Glass Ltd (AIS), PepsiCo, Reebok, Tata Motors and The Oberoi Group.

Najma Khatoon has been a research scholar at The Business School, University of Jammu since 2020. Pursuing PhD in Marketing. Attended workshops and conferences across the country. She has published two papers in reputed journals, one in *European Economic Letters* and another in *Business Strategy and Development*.

Parikshita Khatua, Assistant Professor in KISS DU, Odisha, India, has 20 years of teaching experience in UG and PG level. He has published 20 research articles in both national and international journals and guided research scholars, and two of them have been awarded by the KIIT University. He has published a book and is teaching research methodology, statistics for decision-making to PG ad PhD students. He is also associated with NSS activities as a programme officer in KISS DU.

Tinashe Abraham Kodzanai has BSc and MSc degrees in Tourism and Hospitality Management from the University of Zimbabwe. He is a Principal Lecturer in the Department of Hospitality and Tourism at Harare Polytechnic College, Zimbabwe. He is a member of Tourism Researchers Association of Zimbabwe (TERAZ). His teaching areas include tourism economics, tourism accounting, hospitality computing and travel. His current research interest focuses on women in tourism, rural tourism, community-based tourism and ICT in tourism.

Rajinder Kumar is an Assistant Professor at the Department of Travel and Tourism Management, University of Ladakh, India. For 8 years, Dr Kumar is teaching at the undergraduate and postgraduate levels. His area of research interest is destination performance analysis, tourism education, workations, geotourism and event performance measurement.

Raushan Kumar is currently an Assistant Professor at Amity College of Commerce Haryana, Amity University. He is pursuing his PhD (pre-submission stage) from Mahatma Gandhi Central University, Bihar. He has completed post-graduation from National PG College Lucknow and graduation from Sri Aurobindo College, Delhi University. He has authored more than 10 research papers, articles and edited books in national and international journals and websites. He is a life member of All India Accounting Association. He has experience in many nation-building organisational and development sectors. His area of interest is in finance and financial inclusion.

S. Kumaran is working as a Professor in Alliance School of Business, Alliance University, Bangalore. He has a total of 30 years of experience (14 years in the industry and another 16 years in teaching and research), and he has published five books and more than 60 research papers in journals and conferences. Till now, two PhDs have been awarded under his guidance, and six more scholars are pursuing PhDs under his guidance. His objectives, education, experience and accomplishments, are briefly given for your kind perusal and consideration. Worked as a Business Development Coordinator in Texmo Industries, Coimbatore, and organised more than 500 business development programmes for 1,200 dealers across the country. Worked as a Coordinator for NGO networks in Tamil Nadu, Andhra Pradesh and Karnataka states and facilitated them in planning, funding, monitoring and evaluation of resource management projects. Organised and conducted more than 100 training programmes for developmental activists.

B. Latha Lavanya has 19 years of experience. She has been a Senior Human Resources Manager for about 7 years, full-time PhD (research fellowship awarded by UGC) candidate from the University of Madras. Her accolades include Best Faculty in Management Award in the year 2014 from Education Today, Research Excellence Award in the year 2017 by Indo Global Education and Award for Teaching Excellence by GMRAF Awards in the year 2019. Best women researcher Award of the year 2020. She has been associated with educational institutions for the past 11 years. She has done her postgraduate degree in MHRM and MBA. She holds University of Madras PhD Guide Ship. She is the author of over 50 research papers which were published in various UGC Care and Scopus, Web of Science and ABDC journals. She is currently an Editor for MIMIJMR. She is the first faculty credited to receive a funded internship from Reserve Bank of India and received a grant from ICSSR to conduct a national conference in MIM. Her expertise comprises quantitative techniques, HRD, HRM, innovation, entrepreneurship, logistics and SCM, marketing management, hospital management and knowledge management.

Alka Maheshwari has over 30 years of experience in academics. She had worked for 17 years in South Delhi Polytechnic for Women. Dr Alka started her journey as a Lecturer and finally reached the designation of Head of the Department. In her sojourn, she was fortunate to represent the institute at Lancashire University, Blackburn as a team member for UKIERI. She joined Amity University in 2008 as an Assistant Director, Amity Polytechnic for Women, and is working as an Associate Professor in Amity Institute of Travel and Tourism. She has a Master's degree in English Literature and Tourism, and her PhD is in Community-Based Tourism. She has publications in various national and international journals.

Rekha Maitra is a Seasoned Academician with a Doctor of Philosophy in Tourism and Hospitality (PHDTS). She is a Hotel Management professional, having 23 years of experience with extensive knowledge. Her area of expertise is travel and tourism, housekeeping and front office. Currently, she is working with GD Goenka Public School, East Delhi. Dr Rekha has more than 30 publications to her credit. She has made publications in journals, books and magazines. She is on the Editorial Board of the *International Journal of Social Media and Online Communities* (IJSMOC), the *International Journal of Research in Tourism and Hospitality* (IJRTH) and the *Asian Mirror International Research Journal*. She was associated with several publication houses as a Book Reviewer. Dr Rekha is a recipient of the Best Teacher Award for Travel and Tourism Management by IHMS, Kotdwar (2022) and Indian Hospitality Congress Award, i.e. Aspiring Researcher of Tourism and Hospitality Award (2014).

Bartolomé Marco-Lajara (PhD) is a Professor at the University of Alicante. His research interests are strategic management and tourism management. He is the author of several books, book chapters and international articles related to the areas mentioned above. He is a member of the Tourism Research Institute at the University of Alicante since its foundation and the main researcher of the European Project 'Next Tourism Generation Alliance' at his University, as well as of the public competitive project for the creation of the Tourist Observatory of the Valencian Community (Spain). He has taken part in other public and private projects, such as the development of the strategic plan of the Alicante province for the period 2010–2020. He is the Head of the Department of Management at the University of Alicante and previously was the Assistant Dean of the Economics Faculty for the Business Degree.

Javier Martínez-Falcó (PhD) is an Assistant Professor at the Management Department of the University of Alicante and a Research Fellow at Stellenbosch University. In the field of research, he focuses his interest in the wine business, specifically in the areas of wine tourism and wineries' intellectual capital, on which he has written hundreds of publications in the form of articles, book chapters and contributions to conferences. He has also participated in several national and international research projects, being a Lecturer of Strategic Management in the economics and business degrees of the University of Alicante.

Joaquín Martínez-Falcó holds a degree in History from the University of Alicante. He has a Master's degree in Contemporary History from the

Autonomous University of Madrid and a Master's degree in Secondary Education from the University of Alicante. He has completed a mobility grant at the Andrés Bello University in Chile. He works as a History Teacher in secondary education and is currently a PhD student at the University of Alicante on the relationship between the wine industry and historical heritage.

Livingstone Mhonda is a PhD candidate in Tourism and Hospitality Management with the University of Zimbabwe. He has a BSc degree in Tourism and Hospitality Management (2003) and an MSc degree in Tourism and Hospitality Management (2020). His research involves the interrogation of ecotourism framework to reconcile the rhetoric, narratives and discourse of ecotourism with the practice in Zimbabwe. Mr Mhonda is a founding member of the Tourism Researchers Association of Zimbabwe (TERAZ). He currently works as a Lecturer in the Department of Tourism and Hospitality Management at the University of Zimbabwe, Harare, Zimbabwe.

T. Milton is currently working as Dean, Tourism and Hospitality Management, Bharath Institute of Higher Education & Research, Chennai. He has more than 31 years of experience in the hotel industry as well as hospitality teaching. He completed his PhD in Tourism Management. Dr T. Milton has published 95 articles in Web of Science, Scopus, UGC Care listed journals and authored six books in Tourism and Management topics. He has presented several research papers in international/national conferences and won several prizes in paper presentations. He has received various awards such as the Best Tourism Professor Award in Hotel Management by Society of Hospitality Educators, Outstanding Teacher Award from Research India Foundation and Best Dean Award from International Award Ceremony on Star Achievers in Engineering, Management, Arts and Science (SAEMAS - 2022), Knowledge Research Academy. He has travelled to several countries such as Malaysia, Sri Lanka, Uzbekistan, Myanmar and Singapore.

Jitendra Mohan Mishra is an accomplished academician with 18 years of teaching experience. He currently serves as the Dean of the School of Management Studies and is the Head of the Department of Tourism and Hospitality Management at Central Tribal University of Andhra Pradesh. His contributions include 8 completed and 5 ongoing PhD studies, 60 paper presentations and leadership in major research projects. Prof Mishra's passion for outdoor activities, including trekking and yoga, complements his academic pursuits. He has authored 40 research articles and five books including the one, *Tourism Principles and Practices*, published by Oxford University Press.

Shreeansh Mishra is working as an Assistant Professor of Tourism in Central University of Tamil Nadu, India. His research highlights the transformative role of self-help groups, microfinance and community readiness in advancing community-based tourism and fostering sustainable development. His work examines factors that synergise marginalised communities and foster sustainable development. With a commitment to global collaboration and continuous efforts, Shreeansh Mishra continues his mission to create a more inclusive and equitable tourism landscape, where gender disparities are addressed and women's empowerment takes centre stage, catalysing a brighter future for all.

Liji Panda, Assistant Professor in KISS DU, Odisha, India, was a very good student which led her to receive the award of Gold Medallist in her Bachelor's in Commerce. She has 15 years of experience in teaching and has been actively involved in the presentation of papers in various international and national seminars and conferences. She further has 20 research papers published in various reputed journals to her credit. Her areas of interest include finance and accounting.

Jyoti Prakash, MSc, harmonises 8+ years of industry mastery with academia at The Lalit Suri Hospitality School, Faridabad. An expert in gastronomy and hospitality, holds various publications in UGC, WoS and other related journals and moulds future leaders, dedicated to advancing hospitality management through inventive academic and professional insights.

Anita Rinomhota is a PhD candidate in Economics and Business Management in the faculty of Management and Entrepreneurial Science of Women's University in Africa. She has a BCom Hons degree in Tourism and Hospitality Management (2006) and an MSc degree in Tourism and Hospitality Management (2013). Her research involves the interrogation of an agritourism framework for empowering women. She currently works as a Junior Lecturer in the Department of Tourism and Hospitality Management at the University of Zimbabwe, Harare, Zimbabwe.

Eduardo Sánchez-García (PhD) is an Assistant Professor in the Department of Business Organization at the University of Alicante (Spain). His research is in the field of strategic management of the company. Specifically, he focuses on the study of the territorial agglomeration of companies in clusters and the absorptive capacity, and its influence on the innovative performance and entrepreneurial orientation of companies. He is the co-author of several papers presented at conferences, both national and international, as well as several articles and book chapters related to his field of research.

Aditi Sharma holds a PhD in Management from Panjab University, Chandigarh, and an MBA from Himachal Pradesh University, Shimla. Her doctoral thesis, titled 'An Empirical Study on Organizational Culture and its Impact on the Leadership Styles and Organizational Effectiveness of the Textile Industry in Himachal Pradesh and Punjab', reflects her deep-seated interest in understanding the dynamics of organisations culture and leadership. With over a decade of teaching experience, Dr Sharma has been instrumental in shaping the minds of future business leaders. As an Assistant Professor in HPKV Business School, School of Commerce and Management Studies, Central University of Himachal Pradesh, her career has been marked by a commitment to excellence in both teaching and research.

Ruchi Sharma received her PhD in Management from Himachal Pradesh University Shimla. She is working as an Assistant Professor in HPKV Business School, SCMS, Central University of Himachal Pradesh, India. Her research interest area is consumer behaviour and general management. She has published many research articles in marketing in various national and international journals and presented research papers in the field of consumer behaviour in national and international conferences.

Vipin Singh, PhD, harmonises 12+ years of industry mastery with academia at VIT University. A vanguard in hospitality, celebrated for Scopus-indexed publications and moulding future leaders, dedicated to advancing hospitality management through inventive education and industry insights.

Syafri Tuharea is the Head of the Public Service Agency – Regional Technical Implementation Unit (BLUD UPTD) for Marine Protected Area Management in Raja Ampat, Indonesia. In this role, he supervises the Marine Protected Area (MPA) administration, acting as an extension of the Marine and Fisheries Service (DKP) under the West Papua Provincial Government. His responsibilities include overseeing regulatory compliance within the MPA zones, promoting conservation awareness among the public, fostering sustainable tourism practices and enhancing the economic advantages of conservation initiatives for the local communities.

Providence Warinda holds a Bachelor's degree in Tourism and Hospitality and a Master of Science degree in Social Ecology. She is currently final year PhD student at the University of Pretoria. She is a highly organised and insatiably curious lady with an interdisciplinary approach to research. Ms Warinda enjoys research and has more than 8 years of research experience specialising in creating data collection tools using ODK, Kobo collect and CSPro software. She also has strong qualitative and quantitative data analysis skills using NVivo, Atlas TI, SPSS and STATA. Her research interests are in ecotourism, tourism development, community development, community-based tourism, rural livelihoods and climate change.

Foreword

Prof. Gauri Dutt Sharma
President, AIU, New Delhi
Vice Chancellor, University of Science and Technology

It gives me immense pleasure that a book on 'Sustainable Tourism: A Comprehensive Multidimensional Perspective' has been compiled by Dr Disha Sharma, Dr Pramendra Singh and Dr Hamid Abdullah.

The world of tourism is undergoing a significant transformation, driven by a growing awareness of the need for sustainability and responsible practices. In this evolving landscape, the concept of sustainable tourism has emerged as a crucial focal point, encapsulating the delicate balance between economic progress, environmental conservation and social development. I am sure that book on, 'Sustainable Tourism: A Comprehensive Multidimensional Perspective', probes into the multifaceted dimensions of sustainable tourism, exploring various case studies, research findings and insights from experts across the globe.

The intersection of economic progress and environmental conservation within the context of hosting extensive international events, with a spotlight on Expo 2020 Dubai. Prof Anushka Lydia Issac offers a comprehensive analysis of Expo 2020 Dubai's sustainability framework, showcasing how the event navigates the challenges of sustaining a global showcase while upholding environmental stewardship.

The challenges faced by ecotourism establishments in Zimbabwe provide valuable insights into the factors affecting the potential for sustainable ecotourism in the region. This highlights the importance of effective marketing, collaboration, and community involvement, as well as the economic, social, and environmental aspects of managing tourist influx. The collaboration between stakeholders and conservation initiatives is crucial in this context.

The new transformative aspects of green tourism in the Himalayan region have created profound revolution in the lives of local communities; on the other hand, Spain's wine tourism industry, focusing on economic, social and environmental benefits of wine tourism, is offering insights for industry leaders and policy-makers. The Vibrant Village Programme in Ladakh, India, has significant socio-economic impacts and promotes sustainable tourism. Tourism has played an important role in the economic development of the Jammu and Kashmir region, indicating its potential to drive economic growth and regional advancement.

The book has been divided into 15 chapters. The contributors have been selected both from India and foreign institutes. This gives an idea of fast-growing tourism industry both at national and international levels. It is influencing the economy of the country and bringing new opportunities to the youth.

The comprehensive exploration of sustainable tourism from diverse perspectives – spanning environmental conservation, community development, economic growth, and cultural preservation –makes this book a valuable resource for researchers, policymakers, industry practitioners, and students interested in advancing sustainable tourism practices globally. I congratulate the editors for their committed efforts for editing new knowledge and new approach for sustainable tourism industry creating academic and industry linkages.

Preface

The world of tourism is undergoing a significant transformation, driven by a growing awareness of the need for sustainability and responsible practices. In this evolving landscape, the concept of sustainable tourism has emerged as a crucial focal point, encapsulating the delicate balance between economic progress, environmental conservation and social development. This book, 'Sustainable Tourism: A Comprehensive Multidimensional Perspective', probes into the multifaceted dimensions of sustainable tourism, exploring various case studies, research findings and insights from experts across the globe.

Chapter 1 sets the stage by examining the intersection of economic progress and environmental conservation within the context of hosting extensive international events, with a spotlight on Expo 2020 Dubai. Prof Anushka Lydia Issac offers a comprehensive analysis of Expo 2020 Dubai's sustainability framework, showcasing how the event navigates the challenges of sustaining a global showcase while upholding environmental stewardship.

In Chapter 2, Mr Livingstone Mhonda and colleagues shed light on the challenges faced by ecotourism establishments in Zimbabwe, emphasising the need for a balance between social, economic and environmental goals. The chapter provides valuable insights into the factors affecting the potential for sustainable ecotourism in the region, highlighting the importance of effective marketing, collaboration and community involvement.

Moving on to Chapter 3, Maulita Sari Hani and Syafri Tuharea present a case study of manta ray tourism in Raja Ampat, Indonesia, examining the economic, social and environmental aspects of managing tourist influx. The chapter offers recommendations for sustainable tourism practices, emphasising the importance of collaboration between stakeholders and conservation initiatives.

Chapter 4, authored by Dr Arun Bhatia, Dr Aditi Sharma and Dr Ruchi Sharma, explores the transformative aspects of green tourism in the Himalayan region. The authors delve into the case studies of green entrepreneurs, illustrating how their initiatives have brought about a profound revolution in the lives of local communities.

In Chapter 5, Javier Martínez-Falcó, Eduardo Sánchez-García, Bartolomé Marco-Lajara and Joaquín Martínez-Falcó take us on a journey through Spain's wine tourism industry, focusing on the sustainable practices adopted by Bodegas Franco-Españolas. The chapter highlights the economic, social and environmental benefits of wine tourism, offering insights for industry leaders and policymakers.

Chapter 6, authored by Dr Rekha Maitra, explores the role of green technology in reducing carbon footprints in the tourism industry, focusing on selected hotels in Delhi. The chapter investigates existing sustainable practices and their impact on carbon footprints, highlighting the need for resource optimisation and environmental compliance.

In Chapter 7, Dr Vipin Singh and Jyoti Prakash examine sustainable tourism initiatives for tribal community development in India, emphasising the potential of tourism to drive economic empowerment and environmental sustainability in tribal regions.

Chapter 8, authored by Dr Rajinder Kumar, Dr Hamid Abdullah and Tsering Chusket, presents a case study of the Vibrant Village Programme in Ladakh, India, evaluating its performance and socio-economic impacts. The chapter provides valuable insights for destination management and event planners.

Chapter 9, authored by Dr Farah S. Choudhary and Najma Khatoon, explores the impact of digital content on destination loyalty, focusing on Ladakh tourism. The chapter examines the correlation between social media activity and visitor loyalty, offering insights for tourism marketers and destination managers.

In Chapter 10, Dr S. Kumaran, Dr B. Latha and Dr T. Milton present an analytical study of tourist visits to Kumbakarai Waterfalls, highlighting the importance of tourism facilities and transportation infrastructure in enhancing tourist experiences.

Chapter 11, authored by Kingshuk Biswas, showcases the solarisation of Modhera Sun Temple in Gujarat, India, as a sustainable energy initiative. The chapter explores the integration of solar energy with historical heritage, offering a model for rural electrification and sustainable tourism.

Chapter 12, authored by Mr Raushan Kumar, Mr Niranjan Deo Pathak and Shiv Swaroop Jha, examines the role of tourism in the economic development of the Jammu and Kashmir region, emphasising the potential of tourism to drive economic growth and regional advancement.

Chapter 13, authored by Dr Alka Maheshwari and Dr Aruditya Jasrotia, focuses on the development of ecotourism in Himachal Pradesh, India, with a case study of Spiti Valley. The chapter explores the economic and environmental benefits of ecotourism, highlighting its potential for sustainable tourism development.

In Chapter 14, Parikshita Khatua and Liji Panda critically examine the role of subsidies in promoting sustainable tourism and economic growth. The chapter offers insights into the strengths and weaknesses of subsidy programmes, providing recommendations for policymakers and stakeholders.

Chapter 15, authored by Shreeansh Mishra, Jitendra Mohan Mishra and Vaibhav Bhatt, explores the role of self-help groups in promoting destination sustainability. The chapter examines the impact of Self Help Groups (SHGs) on economic empowerment, cultural preservation and environmental conservation, offering valuable insights for sustainable tourism development.

Collectively, these chapters offer a comprehensive exploration of sustainable tourism from diverse perspectives, spanning environmental conservation, community development, economic growth and cultural preservation. We hope that this book serves as a valuable resource for researchers, policymakers, industry practitioners and students interested in advancing sustainable tourism practices globally.

Acknowledgements

Dr Disha Sharma: I would like to express heartfelt gratitude to my parents, Mr Sharad Sharma and Mrs Bina Sharma, and my husband Dr Ashish Singh, for their unwavering support, understanding and motivation throughout the journey of this book. I would like to dedicate this book to my Head Prof (Dr) Sumita Dave for her constant guidance and motivation.

Dr Hamid Abdullah: I dedicate this book to my late father Mr Ghulam Hussain, my mother Mrs Hasim Bibi, my brother Abdul Rahim and my guide Prof Sunil K. Kabia who have been the force behind me in all my endeavours.

Dr Pramendra Singh: I would like to dedicate this book to my family for their unwavering support and motivation during the creation of this book. I would also like to thank my colleagues and friends for their guidance throughout the editing process.

Chapter 1

Sustainability at the Crossroads of Global Showcases: Balancing Economic Progress and Environmental Conservation in Expo 2020 Dubai

Anushka Lydia Issac[1]

Westford University College, UAE

Abstract

In an era where exceptional exhibitions and mega-events garner global attention, the association between economic advancement and environmental conservation grabs focus. This has never been more apparent than at Expo 2020 Dubai, a global event that incorporated progress, innovation and culture (Haneef & Ansari, 2019). This chapter aims to explore the relationship between sustainability, economic progress and environmental conservation through the lens of Expo 2020 Dubai. The research approach relies on prominent case studies, accentuating the contribution of Expo 2020 Dubai towards the global discussion on sustainable event management. This chapter dives into the sustainability framework, exploring strategies including eco-friendly infrastructure, energy efficiency, waste management and community engagement. It demonstrates Expo 2020 Dubai's proactive approach in addressing the challenges of hosting mega-scale events sustainably. This chapter also adds value by elucidating the parallel sustainability journey, that is otherwise, often overshadowed by the economic aspects of global events (AML Gomba et al., 2018). On the whole, this chapter significantly contributes towards the understanding of sustainable event management by accentuating Expo 2020 Dubai. It highlights the event's role in driving environmental awareness and innovation, demonstrating how such mega-scale events can inspire future sustainable practices (Cull, 2022).

[1]https://orcid.org/0000-0002-1493-1047

Sustainable Tourism, Part B, 1–18

doi:10.1108/978-1-83608-326-920241001

Keywords: Sustainability; economic progress; environmental conservation; Expo 2020 Dubai; global showcases; green initiatives; eco-friendly technologies; Sustainable Development Goals; biodiversity conservation; innovation in sustainability

1. Introduction

> The UAE is striving to develop and boost its rich resources and expertise in the international energy markets and enhance its leading role as a world center for renewable energy research and development. His Highness Sheikh Khalifa bin Zayed Al Nahyan. The President of the UAE (Weatherby et al., 2023).

Travel and tourism, one of the major global economic sectors, boosts exports, produces jobs and spreads wealth globally. The United Nations declared the year 2017 as the International Year of Sustainable Tourism for Development (UNESCO), aimed at offering a chance to highlight the sector's immense potential contributions to the economy, society, culture, environment and heritage year (Haneef et al., 2019). While the idea of tourism as a novel human endeavour with a variety of implications developed gradually, the earliest attempts to characterise it appeared throughout the 19th and 20th century transition (Cunha, 2012). The educational aspects of travel were highlighted by philosophers like Montaigne (1942), prior to this; writers only afterwards focused on the economic benefits of travel. This is the case with the French writer Stendhal (1995), who is credited with coining the term 'touriste', and the Portuguese writer Herculano (1838), who almost simultaneously emphasised the benefits the country gained from foreign visitors who came to indulge in the landscapes or the heritage.

Sustainable tourism, often hailed as the responsible alternative to traditional mass tourism, has garnered significant attention in recent years due to its purported commitment to environmental conservation, social responsibility and economic development (Cull, 2022). The principle behind sustainable tourism is to maximise the positive effects of travel while minimising its detrimental effects on the environment, society and local economies. Sustainable tourism addresses the needs of tourists, the travel industry, the environment and host communities by thoroughly considering the economic, social and environmental effects of travel both now and in the future. It is not an exclusive kind of tourism; rather, all types of tourism could aim to become more environmentally friendly (United Nations Environment Programme (UNEP) & World Tourism Organization (UNWTO), 2005).

The preferred notion of *'Sustainable Development'* proposed by the World Tourism Organisation (WTO):

> Sustainable Tourism Development meets the needs of present tourists and host regions while protecting and enhancing opportunities for the future. It is envisaged as leading to the

> management of fall resources in such a way that economic, social, and aesthetic needs can be fulfilled while maintaining cultural integrity, essential ecological processes, biological diversity, and life support systems.

While the definition of '*alternative*' or '*responsible*' tourism is very broad and occasionally ambiguous (Theobald, 2013), case studies from the past and present that address challenges related to sustainable tourism typically understand '*alternative*' tourism to be the antithesis of mass tourism (Weaver, 2007). Tourism that is categorised as '*alternative*' is thought to be locally managed, thoughtfully planned, qualitative and value conscious. In contrast to traditional tourism, a sustainable approach aims to preserve and improve prospects for future generations while also acknowledging the value that local people, customs and lifestyles provide to the visitor experience. According to Cater (1993), there are three main goals for sustainable tourism: preserving the environment in order to accomplish the first two goals, meeting the needs of the host population in terms of bettering living standards both temporarily and permanently and catering to the needs of an increasing number of tourists. The 'sustainability trinity', which strives for the seamless and transparent integration of the economy, society and environment, is highlighted by Farrell (1999). Liu (2003) claims that sustainable tourism requires both the environmentally friendly use of resources and the sustainable expansion of tourism's contribution to the economy and society. Without a solid understanding and appropriate management of tourism demand, neither can be accomplished.

The theme of this chapter encapsulates the essence of this broader challenge. In a period when the preservation of natural and cultural heritage competes with the relentless advance of contemporary life, the pursuit of equilibrium becomes both a philosophical and practical imperative. Our globe is witnessing a slew of initiatives aimed at striking this equilibrium, but none is arguably more iconic than Expo 2020 Dubai. This worldwide display represents not only the pinnacle of human achievement and creativity but also serves as a litmus test for our commitment to sustainability. Across literature, some authors (Desai, 2013; Henderson, 2006; Ryan et al., 2012; Sangeeta & Anandkumar, 2014; Saxena, 2011; Zaidan & Development, 2016) happen to have characterised tourism growth in a more conventional '*destination*' sense (Butler, 1980; Gunn, 1993), but few have examined the patterns of that growth or, perhaps more importantly, addressed the political commitment (Brent Ritchie & Crouch, 2010) that has both facilitated and fuelled the rapid growth of Dubai as an international tourism hub.

In the broader perspective, this chapter seeks to go beyond the immediate confines of Expo 2020 Dubai. It aims to contribute to a global conversation about sustainable tourism, by investigating strategies that cross geographical boundaries. By critically evaluating the intricate interplay between progress and conservation, this chapter aims to offer nuanced insights and practical solutions.

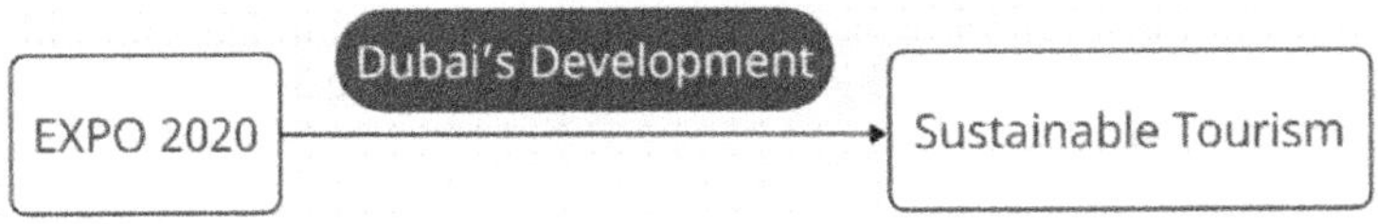

Fig. 1.1. Conceptual Framework. *Source:* Author.

1.1 Conceptual Framework

Within the scope of Expo 2020 Dubai, the conceptual framework (See Fig. 1.1) of this study navigates the complex intersectionality of environmental protection, economic advancement and sustainability. Expo 2020 is the independent variable that is driven by the combined force that exists between *'sustainable'* or *'responsible'* travel and Dubai's continued growth. The dependent variable indicates the extent to which Expo 2020 promotes behaviours that minimise environmental damage while maximising societal and economic advantages. The dynamics between Expo 2020 and sustainable tourism are influenced by Dubai's development, which acts as a moderator variable and the last arm of the framework. This framework probes the opportunities and challenges associated with striking a balance between environmental conservation and economic advancement by delving into the complex dynamics, trade-offs and potential synergies that emerge when sustainability is placed at the centre of a global exhibition.

2. Dubai's Portrayal as a Tourism Destination

Among the United Arab Emirates (UAE)'s seven Sheikhdoms (Emirates) are Dubai, Abu Dhabi, Sharjah, Ajman, umm al-Quwain, Ras al Khaimah and Fujairah. Out of the 4.1 million people living in the United Arab Emirates, 79.9% are non-nationals, and there were approximately 825,000 Emiratis in the 2005 census.

The Dubai Commerce and Tourism Promotion Board (DCTPB) was founded in 1989 with the goal of promoting the destination to business and upscale travellers from predominantly wealthy societies (Laws, 1995). In 1997, the DCTPB was replaced by the Department of Tourism and Commerce Marketing (DTCM), which was granted more authority to manage the development, planning and oversight of the tourism industry. DTCM manages marketing across continents like Africa, Australia, Europe, North America and southern and eastern Asia and runs about 18 overseas offices in important tourist-producing nations. While it's a common belief that Dubai's rise as a global travel destination has occurred in the last 10 years, the early to mid-1990s was a more pivotal time for development as it laid the groundwork for an organised tourism infrastructure. A thriving and alluring services sector industry was made possible by the building industry's expansion in the 1970s and 1980s. Ten international sporting events were held between June and December 1993, and seven airlines started offering direct flights from the United Kingdom to Dubai in January 1994 (Laws, 1995).

Table 1.1. Global Top 10 Destination Cities by International Overnight Visitors, 2018.

Rank	Destination City	2017	2018	2019 (Growth Rate Forecast)
1	Bangkok (Thailand)	21.09	22.78	3.34%
2	Paris (France)	17.41	19.10	2.24%
3	London (United Kingdom)	19.83	19.09	3.47%
4	Dubai (UAE)	15.79	15.93	1.68%
5	Singapore (Singapore)	13.90	14.67	4.00%
6	Kuala Lumpur (Malaysia)	12.58	13.79	9.87%
7	New York (United States)	13.13	13.60	2.94%
8	Istanbul (Turkey)	10.70	13.40	8.14%
9	Tokyo (Japan)	11.93	12.93	10.02%
10	Antalya (Turkey)	9.42	12.41	8.14%

Source: Adapted from Statista.com: The statistical portal, www.statista.com/statistics/284636/visitorarrivals-in-Dubai-from-international-destinations

From about 374,000 arrivals in 1982 (Henderson, 2006) to just over 15 million in 2017 (Statista.com: The statistical Portal) (As indicated in Table 1.1), the destination has evolved into a 'lifestyle' destination known for sun, sea, sand and designer shopping, in addition to some of the most opulent hotels and golf courses in the world. In contrast to the other Gulf countries, Steiner (2009) emphasises that '*Dubai is the undisputed epicenter of an iconographic destination development*' in which Dubai is a '*hyperreal*' destination rather than a historical or cultural destination, with growth powered by the creation of new hotels and attractions designed to cater to Western tourists. Anthonisz and Mason (2019) also discuss the hyperreality of the service experience and the '*theme park*' approach to strategy that looks to propel the Dubai brand ahead in terms of success. Initially, the DTCM, which replaced the Dubai Commerce and Tourism Promotion Board in 1997, and Emirates Airlines, which was founded in 1985, were the two main amplifiers of the Dubai brand's development and its establishment as a '*hyperreal*' destination. The visionary leadership of His Highness, Sheikh Mohammed bin Rashid Al Maktoum (DTCM), the city's advantageous location, its reputation for safety and its multi-award-winning marketing campaigns have all contributed to the development of Dubai's tourism strategy (Anthonisz & Mason, 2019). The rise in tourism corresponds with the expansion of the national flag carrier, Emirates Airlines, which has made use of Dubai's strategic location to expand globally and now flies to 154 destinations in 86 countries (The Emirates Group).

According to the Dubai Economic Report (Government of Dubai, 2018), there are 20 different economic sectors that make up the varied Dubai gross domestic product (GDP). The tourism industry, which includes lodging and dining, saw the most gain. The industry with the fastest growth rate was tourism. With an 8% GDP growth rate in tourism, the sector was a major driver of Dubai's overall GDP growth rate in 2017. In 2019, UAE Tourism was ranked (33rd) internationally and first in the Middle East (The World Economic Forum, 2019). According to the 2019 Global Destination Cities Index survey (Mastercard, 2019), the Emirate saw 15.93 million foreign overnight visitors in 2018 and is predicted to continue growing on this success in 2019. Among the world's most visited cities is Dubai. For the fifth consecutive year, Dubai has remained the fourth most visited city worldwide.

In terms of geography, Dubai's desert environment remains incompatible with large-scale tourist development, given the extreme temperature and likely scarcity of resources to entice visitors. This, however, has not decelerated development in any manner and on the contrary appears to have served as a robust engine towards growth, since Dubai's expansion has not been at the price of displaced inhabitants or constrained by existing infrastructure (Anthonisz & Mason, 2019). Although the emirate's early success was primarily based on trade and income from the acknowledged declining oil reserves, this has enabled the UAE to evade the traditionally lengthy and challenging process of conserving and capital accumulation that accompanies economic growth and rather jump straight to the stage of high mass consumption (Shihab). As a result, Dubai has switched its focus from oil and trade to transforming into one of the world's fastest growing tourism destinations.

It's interesting to note that Dubai's reputation has been deliberately and purposefully built on a platform of five-star (or seven-star) luxury offered by an international brand presence that is tailored to visitor experiences in an Arabic, tolerant and cosmopolitan environment, rather than on offering culturally Islamic tourism experiences for visitors from the surrounding area (Anthonisz et al., 2015). The country's premier travel and business destination is Dubai, which boasts luxurious resorts, championship golf courses, exemplary shopping, prestigious restaurants and renowned architectural sites. According to AML Gomba et al. (2018), a majority of respondents cited 'leisure and entertainment' as their prime reason to visit Dubai, while almost 50% of tourists claimed to be in Dubai for business, and roughly 33% were there to visit friends and family. Further, almost one-third of those surveyed said that Dubai's culture and tradition were what drew them to the city. Fig. 1.2 illustrates the respondents' perception of Dubai as a tourist destination – Global Results, according to a survey conducted by the Centre for Sustainability Through Research and Education – approximately 80% (out of 1,790) of the participants expressed agreement with the 18 sustainability-related statements, suggesting that the majority of them view Dubai as a destination for sustainable tourism.

Scholars typically view Dubai as a strong model of tourism development, even though they acknowledge certain obstacles (Bagaeen, 2007; Govers & Go, 2009; Henderson, 2006; Sharpley, 2008). Bagaeen (2007) highlights Dubai's status as an '*instant city*' and the '*outcome of a super-fast urbanism*' while also acknowledging that, in light of socially divided lifestyles and ongoing environmental concerns, the city may not be able to sustain itself in the long run. According to Govers and

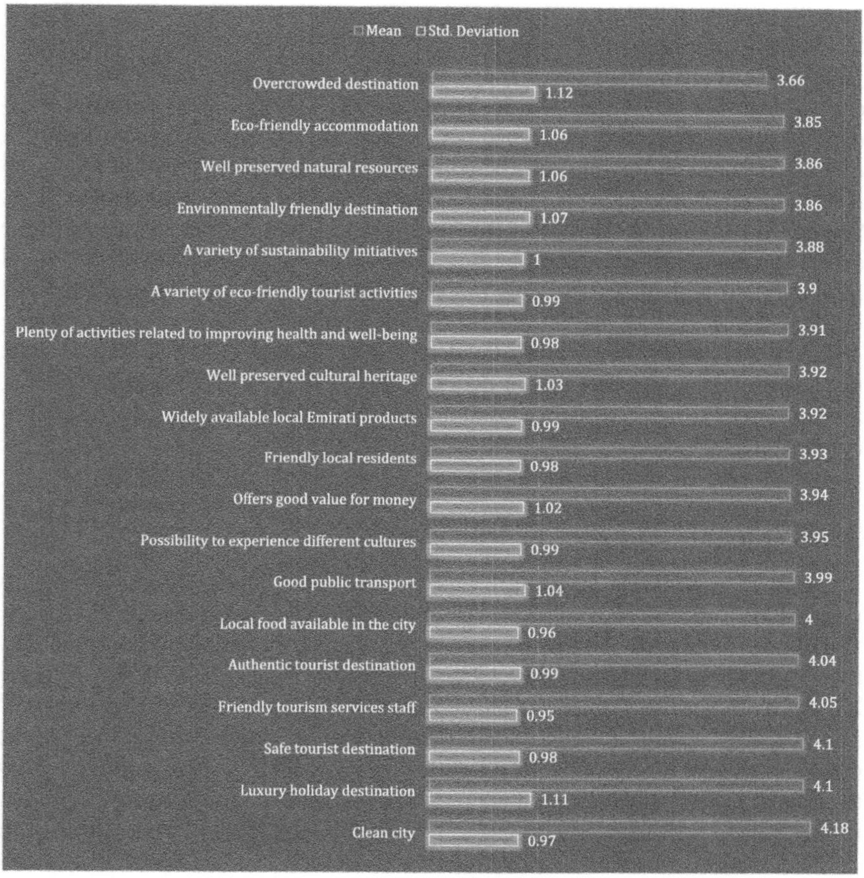

Fig. 1.2. The Respondents' Perception of Dubai as a Tourist Destination – Global Results. *Source:* Adapted from chrome-extension://efaidnbmnnnibpcajpcglclefindmkaj/https://www.globalsurvey-sdgs.com/wp-content/uploads/2020/01/20200205_SC_Global_Survey_Result-Report_english_final.pdf

Go (2009), among other things, the creation and promotion of a wide range of tourism products, creative megaprojects, useful public sector activities and the building of brand equity are all contributing factors to Dubai's status as a '*very powerful international brand*'. According to Sharpley (Govers & Go, 2009), there are a number of proactive 'pro-tourism' policies that can be implemented to boost the growth of the industry. These include international investment, the liberalisation of visa requirements, the revision of restrictions on foreign land ownership for specific tourism projects, non-interventionist trade policies (like '*open skies*') and the easing of stringent moral codes of conduct related to dress code and alcohol consumption.

The micro-reconstruction of the world itself symbolically endorses the conceptual significance of the '*World*' as a symbolic fabrication of Dubai's development ethos. Projecting a science fiction fantasy, the city has given rise to a number of home-grown brands, including the largest shopping mall in the world (Dubai Mall), the largest prize money in the world for a horse race between M. L. Stephenson and J. Ali-Knight (Dubai Cup), the first purpose-built maritime centre (Dubai Maritime City), the first (informally recognised) '*seven-star*' hotel (Burj Al Arab), the tallest building in the world (Burj Kalifa) and the tallest hotel in the world (Rotana Rose Towe Nestled among luxury residences and corporate offices, Burj Kalifa is home to a hotel designed by Giorgio Armani and boasts the fastest lift in the world. Located in the Gulf, five kilometres off the coast of Dubai, are over 300 miniature real estate islands, one for each continent and one for each country inside each.

3. Sustainable Tourism in Dubai

The case of Sustainable Tourism in Dubai is intriguing and complex, with juxtaposition between luxurious technology and the delicate desert nature. Positively, Dubai has established itself as a sustainable travel destination and has made significant progress in supporting eco-friendly activities (Cull, 2022). Along with the Dubai Green Tourism Award, which will be in conjunction with World Environment Day, DTCM has also announced the official launch of the first-ever project to cut carbon emissions (CO_2) in the hotel industry by 20%. The initiative's primary goals are to raise public knowledge of climate change, CO_2 emissions, energy consumption in the tourism sector and environmental practices in the hospitality sector (Ayeni, 2013). Dubai's commitment to environmental conservation is evident in its ambitious projects, such as the Dubai Sustainable City (AL-Dabbagh, 2022; El-Bana, 2015; El-Jisr, 2017; Passerini & Ricci, 2020) and the Green Building Regulations (The Municipality mandated the standards for all new buildings post the construction of 44 green buildings in 2014, to address ecology, planning, vitality, energy, water, materials and waste building design) (Abu-Hijleh & Jaheen, 2019; Salama & AlSaber, 2014, pp. 1309–1320). These initiatives aim to minimise carbon footprints through the integration of renewable energy sources, energy-efficient buildings and extensive green spaces. The city's efforts to implement sustainable infrastructure, including the development of sustainable transportation systems and waste management practices, deserve acknowledgement (Cull, 2022).

4. Expo 2020

Expo, as defined by the Bureau International des Expositions (BIE), is a global event that aims to educate the public, share innovation, encourage progress and foster collaboration. It serves as a global forum for dialogue, bringing people from all over the world together to address the most pressing issues facing humanity. Mega-events boost economic growth, draw attention from around the world (Kim et al., 2012) and influence travellers' perceptions of a place both before and after (Kim & Morrsion, 2005). A mega-event can serve as a catalyst for significant urban redevelopments that are more closely aligned with the goals of attaining a sustainable environment and the economic development of the hosting city. Thus, hosting a mega-event is a great opportunity for the host city to foster development, generate employment, improve the global business and investment environment and develop tourism (Biermann et al., 2017; Grix, 2013; Hiller, 2000; Smith, 2012).

A massive event – the World Expo began as a means of exhibiting the scientific and industrial advancements of the era and is held every five years. Modern Expos also offer a forum for global discussion and cooperation leading to settlement of the main issues confronting society. For instance, strict emission control laws during Expo 2010 led to a significant improvement in Shanghai's air quality. Though sadly, this effect was only temporary, the Shanghai locals perceived the Expo as stimulating communal involvement (Huang et al., 2013). Millions of people attend the event as well; the Shanghai Expo brought in 73 million visitors and generated an 8–10 billion USD profit (Kim et al., 2012). While marketing campaigns have an ability to alter the perception of a destination (Gibson et al., 2008; Gripsrud et al., 2010; Kim & Morrsion, 2005; Lorde et al., 2011), they also tend to have a positive immediate and long-term effect on the host nation. Greece's GDP increased from 1997 to 2005 as a result of the 2004 Olympic Games (Kim et al., 2012). Full-time employment and economic profit both rose as a result of the FIFA 2002 World Cup (Kim et al., 2012). Additionally, this incident enhanced traveller perceptions of South Korea (Kim & Morrsion, 2005). According to Kim and Chalip (Kim & Chalip, 2004), these kinds of mega-events could create opportunities for the travel and tourism sector in the destination.

From 20 October 2020 to 10 April 2021, Dubai, United Arab Emirates (UAE), hosted Expo 2020 – the first World Expo to be held in the Middle East, North Africa and South Asia (MENASA), with an estimated 25 million visitors, 190 participating countries and 94,400 job opportunities, of which 51,100 were in the tourism industry. Positive effects on Dubai were evident since the hosting of Expo 2020 including a significant increase in economic growth, investments, hospitality and tourism and job opportunities across industries, including aviation, real estate and construction, engineering and infrastructure, logistics and transportation and retail (Expo2020). As the tourism industry of Dubai continues to garner attention, certain commentators on tourism (Menon & Somashekharan, 2017; Stephenson & Ali-Knight, 2010) work to advance the politically acceptable notion of a morally and socially responsible industry.

5. Case Study: Interlinking Sustainability With Dubai's Expo 2020

The 2030 Agenda for Sustainable Development, which is supported by several additional international frameworks, is at the centre of this. Ever since the United

Arab Emirates (UAE) submitted its proposal to host Expo 2020, the UN and its member states have constantly recognised global concerns that have influenced the definition, rationale and development of the bid's fundamental topic.

A World Expo is a collaboration between diverse stakeholders. Expo 2020 Dubai offered a forum to discuss numerous facets of the Sustainable Development Goals (SDGs), with a particular focus on areas where the international community ought to concentrate its efforts, through its subthemes. Together with other local and federal organisations, the UAE National Committee on SDGs and other UAE organisations, Expo 2020 Dubai aimed to address the 2030 Agenda for Sustainable Development and the SDGs (National Committee on Sustainable Development Goals UAE, 2017). Through the use of advocacy and communications, policy discussions and creative solutions for successful implementation, this assisted in bringing together pertinent thought leaders and practitioners to deliver a multi-stakeholder approach that allowed for engagement on all levels and across all Expo 2020 Dubai initiatives. Expo 2020 proposed and followed through a sustainable strategy to reuse 80% of the infrastructure, buildings and facilities built for the event for legacy purposes. After Expo 2020 wrapped up, post its six-month event time, the Dubai Expo site has pushed the limits of sustainability, smart technology, architecture and accessible transportation.

Facilitating a global conversation at Expo 2020 is one of the main objectives of the masterplan. It successfully accomplished this by bringing the main subject, '*Connecting Minds, Creating the Future*', to life and skilfully combined the three subthemes of opportunity, mobility and sustainability. The 4.38 km^2 Expo 2020 site is home to 180 national pavilions. Al Wasal Plaza, also known as '*The Connection*', is a massive dome located in the centre of Expo 2020 that unifies the several subthemes into a unified, expansive area that physically embodies the exhibition's central subject (EXPO2020 Dubai). The mobility pavilion facilitated the movement of people, commodities and ideas in a more intelligent and efficient manner, so enabling individuals and communities to realise their full potential. Its objective was to investigate how movement – both digital and physical – can serve as a conduit for opportunity – everything from the restructuring of international supply chains to the vital function of infrastructure and human mobility due to migration or tourism, increased knowledge and human mobility propels advancement (National Committee on Sustainable Development Goals UAE, 2017). This subject provided a fresh perspective from which to develop answers, or it is essential to many of the SDGs. Expo 2020 Dubai's subtheme of Mobility included as its main focus the following areas: digital connectivity, personal mobility, travel and exploration, logistics and transportation. Many of the 17 SDGs are related to these important subjects.

On the other hand, the sustainability pavilion promoted eco-friendly living practices that respect environmental constraints. The planned footprint of the Sustainability Pavilion was 29,000 m^2. By utilising solar energy, the pavilion achieved net-zero energy. In order to reach significant water usage goals, humidity harvesting equipment was employed. The Sustainability Pavilion's legacy is that it is designed to transform into a Science Exploratorium with a focus on empowering and inspiring young people to become ambassadors for improved environmental stewardship. Sustainability's subtheme pertains to living in harmony and with the environment we live in. The need to live within the limits of our planet will become increasingly tied to the opportunities we provide for present and future generations in the years to come (National Committee on Sustainable Development Goals UAE, 2017). There is a great deal of space for innovation as our population and cities expand to make sure that growth is sustainable and within the bounds of the planet, that gaps are filled, and that we build the social and physical environments necessary for happier and healthier societies. Expo 2020s Sustainability subtheme covered the following important subjects: mitigating climate change, managing natural resources, preserving biodiversity and natural ecosystems and creating sustainably built environments.

The opportunity pavilion represented Expo 2020 Dubai's pledge to help people realise their full potential so they can build better futures. The Opportunity subtheme relates to encouraging people and communities to realise their potential and influence the course of history. It aimed at addressing the differences that exist in the globe, the importance of developing human capital and the significant changes that are currently taking place in the fields of economic development, employment and education. Expo 2020 Dubai had chosen the following important subjects under the subtheme of Opportunity: new industries, employment opportunities, financial capital, education and governance. Particular areas of activity within each of these major themes that are closely linked to the objectives and the SDGs make it possible for Expo 2020 Dubai to link directly and significantly and transcend beyond a conceptual linkage. For instance, the work on the topic of education encompasses the following areas: online learning and new tools for learning (SDG 4.b), the school of the future (classrooms, tools, programmes) (SDG 4.a), mentoring and support networks (SDG 4.8, 4.9), quality of education (SDG 4.4, 4, 7) and access to education (SDG 4.1, 4.2, 4.3, 4.5). Expo 2020 Dubai intentionally aims to go beyond the agreed targets of the SDGs and re-imagine the goals at their highest potential, free from political constraints, while keeping in mind the importance of delivering these goals to everyone, in a sustainable, equitable and inclusive manner.

These Expo's main themes are directly related to the goals of individual SDGs, including SDG 13 (Climate Action), SDG 11 (Sustainable Cities and Communities) and SDG 7 (Affordable and Clean Energy).

The Expo site had sourced at least half of its energy from renewable sources. About 100 MW of electricity was supplied by the Mohammed Bin Rashid Al Maktoum Solar Park. The information gathered from reports and interviews indicated that the main structure of the event's design and operation provides a significant explanation for the concepts surrounding sustainable planning in leaving a lasting legacy. The idea of the legacy is described as a change from a fantastic world expo to a community that is unlike any other, where people can live, work and conduct business while surrounded by amazing architecture and innovative design (Expo2020; EXPO2020 Dubai).

6. Dubai's Development as a Catalyst Towards Driving Sustainability at Expo 2020

Dubai has grown in importance as a regional and international travel destination, becoming well-known for its shopping, entertainment, athletic events, international conferences and media events. Apart from its robust services industry, Dubai has constructed cutting-edge amenities and tourism infrastructure. All things considered, Dubai's success provides insights into how a state with a dearth of traditional natural and cultural attractions has become one of the most sought-after travel destinations worldwide (Haneef & Ansari, 2019). Furthermore, the development of tourist infrastructure and services frequently has destination attractions as its catalytic focus, making them a complicated area of the tourism industry (Alan, 2008). The population has grown at an equally rapid rate, from 370,000 in 1973 (Dubai Statistics Centre, 2009) to little over 2.5 million in 2016 (World Population Review, 2024). This boom in tourist numbers has coincided with an equally significant increase in population. The primary cause of this population growth is the influx of foreign laborers, who are traditionally brought in to cover tasks that cannot be filled by local laborers and are necessary to support the destination's expansion (Forstenlecher & Rutledge, 2011). As a result, Dubai is currently among the cities with the fastest rates of growth in the world.

In order to guarantee that the Expo's themes – Opportunity, Mobility and Sustainability – are delivered under the motto of 'Connecting Minds, Creating Futures', the authors believe that the destination's hosting of Expo 2020, the only mega-event in the business world (Getz, 2012), served as a 'tipping point' as it developed (EXPO2020 Dubai). A large investment in infrastructure and attractions was necessary to ensure the Expo's success. Building a destination brand with a powerful and representative image remains a constant concern as the tourism industry grows significantly (Pereira et al., 2019). One of the fundamental elements that contributes to UAE's excellent brand ranking is special events. Meetings, Incentives, Conferences and Exhibitions (MICE) tourism has been based in the Emirate of Dubai because it organises a variety of events all year long, including trade exhibits, art exhibitions, musical performances, literature festivals and economic and sporting activities. The Dubai World Trade Centre was founded in 1979 and is the site of significant international events like the Arabian Travel Market, which is regarded as a turning point in the history of travel and hospitality exhibits (Vij et al., 2019).

It is evident that Dubai's development has nearly complied with the stages that Butler's (1980) Tourist Area Life Cycle (TALC) model postulated. However, the city's rapid expansion in terms of both visitor numbers and auxiliary infrastructure has thwarted the associated periods of reflection and consolidation that accompanying the establishment of a sustainable destination. When Dubai's expansion is compared to Butler's model, it offers a thoughtful perspective on the destination's evolution and, to some extent, makes it feasible to make significant inferences about potential futures for the area beyond 2020, when the Expo event was hosted (EXPO2020 Dubai).

7. EXPO 2020 Dubai's Effects on the Travel and Tourism Sector

Dubai, the Expo 2020 host city, is unquestionably a prospective tourist destination due to its strategic location and distinctive natural features, not just in the Middle East but also worldwide. The success of any newly developed tourism offering depends on its capacity to draw tourists and encourage their participation. Each visitor to an attraction goes through a specific experience (attractions of Dubai article).

There are two ways to categorise the effects of Expo 2020 on tourism in Dubai: direct and indirect effects:

Indirect Effects: World-class attractions including the establishment of Crocodile Park, traditional souks, a fish market and a safari park were exclusively designed for Dubai Expo. With the expansion of Dubai International Airport's facilities and capacities to manage the anticipated increase during Expo 2020, travel and tourism was significantly boosted.

Direct Effects: The following were the projections made according to 'The Economic Impact of Expo 2020 Dubai' (Ernst & Young Middle East, 2019): 190 countries were expected to participate at Expo 2020 Dubai, with an expected 25 million visitors.

The predicted total economic impact of Expo 2020 Dubai from November 2013 to December 2031 is as follows:

- From November 2013 to December 2031, Expo 2020 Dubai is expected to generate AED 122.6 billion in Gross Value Added (GVA) for the UAE.
- From 2013 to 31 (or roughly 49,700 FTE jobs year) in the UAE, Expo 2020 is also anticipated to generate 905,200 full-time equivalent (FTE) job-years in the UAE.

From May 2021 to December 2031, the following is anticipated to be the legacy contribution:

- AED 62.2 billion is the expected GVA impact.
- A total of 548,300 FTE job-years, or roughly 53,800 FTE jobs annually, were created in the UAE during this time.

There could potentially be a number of significant effects throughout the legacy era. They are anticipated to have a long-term impact on the UAE economy. A few of these are as follows (Ernst & Young Middle East, 2019):

- Better trade ties between nations as a result of connections made during the Expo.
- Better firm-to-firm business ties as a result of the connections made during the Expo.
- A higher profile internationally for Dubai and the UAE as a place to work, invest and conduct business.

- A rise in the stature and allure of Dubai and the UAE as travel destinations worldwide.

Overall, the hosting of a mega-event like Expo 2020 can accomplish two significant goals, according to Emirates (2020):

- The establishment of new infrastructure to drive longer term economic growth in the host city or to expedite the execution of infrastructure projects that might have taken longer to complete without such an event.
- To enhance the host city's reputation on the international scene by presenting its advantages for investment, tourism and business, as well as its capacity to organise, oversee and carry out a major international event. After hosting such massive events, a number of cities left feeling burdened, heavily indebted and regretting the construction of numerous pointless structures.

In the modern world, the convergence point of environmental preservation and economic progress has emerged as an essential subject of discussion internationally. The conundrum of sustainable development calls attention to the fine line that must be drawn between fostering economic growth and preserving the ecological integrity of our world, particularly in light of international events like Expo 2020 Dubai. It is becoming increasingly important that we navigate this difficult terrain as we find ourselves at a crossroads in history.

8. Conclusion

This chapter takes its chance at exploring the unique relationship between variables such as sustainability, economic advancement and environmental preservation, with emphasis on the international event Expo 2020 Dubai. It further examines the sustainable measures taken throughout the event, specialising on waste management, energy efficiency, environmentally friendly infrastructure and community involvement. This chapter illustrates how Expo 2020 in Dubai complies with the 2030 Agenda for SDGs, reinforcing the connection between sustainability and the event.

The subthemes of opportunity, mobility and sustainability at Expo 2020 are examined in detail, demonstrating the event's commitment towards tackling global issues. The report emphasises the efforts made by Expo 2020 Dubai to leave a lasting legacy by placing a focus on smart technology, accessible transportation and sustainability. It discusses Dubai's progress as a driving force behind Expo 2020 sustainability, offering insights into how the mega-event's hosting served as a turning point in the destination's history. In the framework of sustainable development, Dubai's explosive rise as a popular tourist destination, its infrastructure investments and its emphasis on major events like Expo 2020 are studied. Under consideration is the direct and indirect consequences of Expo 2020 on the travel and tourist industry, including economic contributions, employment generation and potential long-term benefits for Dubai's international profile.

Eventually, by highlighting Expo 2020 Dubai as a noteworthy example, this chapter adds to the global conversation on sustainable event management. It highlights the crucial role of striking a balance between environmental preservation and economic advancement and how major events like Expo 2020 have the power to bring about positive change. For a sustainable and equitable future, navigating the issues that lie at the junction of ecological preservation and economic development is essential.

References

Abu-Hijleh, B., & Jaheen, N. (2019). Energy and economic impact of the new Dubai municipality green building regulations and potential upgrades of the regulations. *Energy Strategy Reviews*, *24*, 51–67. https://doi.org/10.1016/j.esr.2019.01.004

AL-Dabbagh, R. (2022). Dubai, the sustainable, smart city. *Renewable Energy and Environmental Sustainability*, *7*, 3. https://doi.org/10.1051/rees/2021049

Alan, F. (2008). *Managing visitor attractions, business & economics*. Routledge.

AML Gomba, C., Hyland, L., & Morris Paris, C. (2018). Tourist Perceptions of Dubai and Expo2020. *e-Review of Tourism Research (eRTR)*, *15*, 4–5.

Anthonisz, A., Heap, T., & Rawlinson, S. (2015). Re-positioning the authenticity of the five-star service experience in Dubai. In *CHME Conference*.

Anthonisz, A., & Mason, G. (2019). Reinventing tourism: The Dubai phenomenon. *Worldwide Hospitality and Tourism Themes*, *11*, 279–286. https://doi.org/10.1108/WHATT-01-2019-0004

Ayeni, D. A. (2013). Sustainable Tourism Development: A Viable Option to Boost the Nigerian Economy. *International Journal of Education and Research*, *1*.

Bagaeen, S. (2007). Brand Dubai: The instant city; or the instantly recognizable city. *International Planning Studies*, *12*, 173–197.

Biermann, F., Kanie, N., & Kim, R. E. (2017). Global governance by goal-setting: The novel approach of the UN Sustainable Development Goals. *Current Opinion in Environmental Sustainability*, *26–27*, 26–31. https://doi.org/10.1016/j.cosust.2017.01.010

Brent Ritchie, J. R., & Crouch, G. I. (2010). *A model of destination competitiveness/sustainability: Brazilian perspectives* (Vol. 44, pp. 1049–1066). Reservation de Administracao Publica.

Butler, R. W. (1980). The concept of a tourist area cycle of evolution: Implications for management of resources. *Canadian Geographies/Géographies canadiennes*, *24*, 5–12. https://doi.org/10.1111/j.1541-0064.1980.tb00970.x

Cater, E. (1993). Ecotourism in the Third World: Problems for sustainable tourism development. *Tourism Management*, *14*, 85–90. https://doi.org/10.1016/0261-5177(93)90040-R

Cull, N. J. (2022). The greatest show on earth? Considering expo 2020, Dubai. *Place Branding and Public Diplomacy*, *18*, 49–51. https://doi.org/10.1057/s41254-022-00267-1

Cunha, L. (2012). *The definition and scope of tourism: A necessary inquiry*.

Department of Tourism and Commerce Marketing (DTCM). Hotels | Regulation No 1 of 2002. https://www.dubaitourism.gov.ae/en/legislative-news/hotels-regulation-no-1-of-2002

Desai, S. S. (2013). The Dubai shopping festival: A study of visitor's perceptions. *Journal of Marketing Vistas*, *3*, 67–74.

Dubai Statistics Centre. (2009). *Population Bulletin Emirate of Dubai.*

El-Bana, N. (2015). Optimising sustainability at an urban level: A case study of Dubai Sustainable City. In *Sustainable development and planning VII* (pp. 985–995). WIT Press.

El-Jisr, K. (2017). Dubai sustainable city: Shaping future cities. www.thesustainablecity.ae

Emirates, N. B. D. (2020). *Emirates NBD sustainability report.*

Ernst and Young Middle East. (2019). *The economic impact of Expo 2020 Dubai.*

EXPO2020 Dubai *Annual Review*. https://ar2016.expo2020dubai.com/assets/pdf/English/FULL%20REPORT/Expo2020_Annual_Review_2016.pdf

Expo2020: Expo 2020 Dubai story. https://www.expo2020dubai.com/en/understanding-expo/world-expos-history/expo-2020-dubai-story

Farrell, B. H. (1999). Conventional or Sustainable Tourism? No room for choice. *Tourism Management*, *20*, 189–191.

Forstenlecher, I., & Rutledge, E. J. (2011). The GCC's demographic imbalance: Perceptions, realities and policy options. *Middle East Policy*, *18*, 54–67.

Getz, D. (2012). *Event studies: Theory, research and policy for planned events.* Routledge.

Gibson, H. J., Qi, C. X., & Zhang, J. J. (2008). Destination image and intent to visit China and the 2008 Beijing Olympic Games. *Journal of Sport Management*, *22*, 427–450.

Government of Dubai, D. E. (2018). *Dubai economic report 2018.*

Govers, R., & Go, F. (2009). *Place branding: Glocal, virtual and physical identities, constructed, imagined and experienced.* Macmillan.

Gripsrud, G., Nes, E. B., & Olsson, U. H. (2010). Effects of hosting a mega-spot event on country image. *Event Management*, *14*(3), 193–204.

Grix, J. (2013). Sport politics and the Olympics. *Political Studies Review*, *11*, 15–25. https://doi.org/10.1111/1478-9302.12001

Gunn, C. A. (1993). *Tourism Planning: Basics, Concepts and Cases.* Taylor & Francis.

Haneef, S. K., & Ansari, Z. (2019). Marketing strategies of Expo 2020 Dubai: A comprehensive study. *Worldwide Hospitality and Tourism Themes*, *11*, 287–297. https://doi.org/10.1108/WHATT-11-2018-0071

Haneef, S. K., Ansari, Z., & Bhavani, G. (2019). Attractions of Dubai and Expo 2020: An exploratory study. *Worldwide Hospitality and Tourism Themes*, *11*, 266–278. https://doi.org/10.1108/WHATT-01-2019-0007

Henderson, J. C. (2006). Tourism in Dubai: Overcoming barriers to destination development. *International Journal of Tourism Research*, *8*, 87–99.

Herculano, A. (1838). *Para uma Associação de Defesa do Património.*

Hiller, H. H. (2000). Mega-events, urban boosterism and growth strategies: An analysis of the objectives and legitimations of the Cape Town 2004 Olympic Bid. *International Journal of Urban and Regional Research*, *24*, 449–458. https://doi.org/10.1111/1468-2427.00256

Huang, K., Zhuang, G., Lin, Y., Wang, Q., Fu, J. S., Fu, Q., & Deng, C. (2013). How to improve the air quality over megacities in China: Pollution characterization and source analysis in Shanghai before, during, and after the 2010 World Expo. *Atmospheric Chemistry and Physics*, *13*, 5927–5942.

Kim, S., Ao, Y., Lee, H., & Pan, S. (2012). A study of motivations and the image of Shanghai as perceived by foreign tourists at the Shanghai EXPO. *Journal of Convention & Event Tourism, 13*, 48–73.

Kim, N.-S., & Chalip, L. (2004). Why travel to the FIFA World Cup? Effects of motives, background, interest, and constraints. *Tourism Management, 25*, 695–707. https://doi.org/10.1016/j.tourman.2003.08.011

Kim, S. S., & Morrsion, A. M. (2005). Change of images of South Korea among foreign tourists after the 2002 FIFA World Cup. *Tourism Management, 26*, 233–247. https://doi.org/10.1016/j.tourman.2003.11.003

Laws, E. (1995). *Tourist destination management: Issues, analysis and policies.* Routledge.

Liu, Z. (2003). Sustainable tourism development: A critique. *Journal of Sustainable Tourism, 11*, 459–475. https://doi.org/10.1080/09669580308667216

Lorde, T., Greenidge, D., & Devonish, D. (2011). Local residents' perceptions of the impacts of the ICC Cricket World Cup 2007 on Barbados: Comparisons of pre- and post-games. *Tourism Management, 32*, 349–356. https://doi.org/10.1016/j.tourman.2010.03.004

Mastercard. (2019). *Global destination cities index survey.*

Menon, G., & Somashekharan, K. (2017). Sustainable tourism – Dubai roams with responsibility. *Imperial Journal of Interdisciplinary Research, 3*, 817–828.

Montaigne, M. de (1942). *Journal de Voyages en Italie par la Suisse et L'Allemagne en 1580 et 1581.*

National Committee on Sustainable Development Goals UAE. (2017). *Goals UAE and the 2030 agenda for Sustainable development excellence in implementation.*

Passerini, G., & Ricci, S. (Eds.). (2020), *The sustainable city XIV.*

Pereira, L. A., Limberger, P. F., Flores, L. C., & Pereira, M. (2019). An empirical investigation of destination branding: The case of the city of Rio de Janeiro, Brazil. *Sustainability, 11.*

Ryan, C., Ninov, I., & Aziz, H. (2012). Ras Al khor – Eco-tourism in constructed wetlands: Post modernity in the modernity of the Dubai landscape. *Tourism Management Perspectives, 4*, 185–197.

Salama, M., & AlSaber, T. (2014). *Imposing green building regulations in Dubai.* Proceedings 29th Annual Association of Researchers in Construction Management Conference, ARCOM 2013.

Sangeeta, P., & Anandkumar, V. (2014). Dubai shopping festival: Tourists' nationality and travel motives. *International Journal of Event and Festival Management, 5*, 116–131.

Saxena, R. P. (2011). *Dubai mall: A multipurpose destination in the Middle East* (Vol. 1, pp. 1–15). Emerald Emerging Markets Case Studies.

Sharpley, R. (2008). Planning for tourism: The case of Dubai. *Tourism and Hospitality Planning & Development, 5*, 13–30.

Shihab: *Economic development in the UAE.* www.uaeinteract.com/uaeint_misc/pdf/perspectives/12.pdf

Smith, A. (2012). *Events and urban regeneration: the strategic use of events to revitalise cities.* Routledge.

Statista.com: The statistical portal. www.statista.com/statistics/284636/visitorarrivals-in-dubai-from-international-destinations/

Steiner, C. (2009). From heritage to hyperreality? Prospects for tourism development in the Middle East between Petra and the Palm. *Journal of Islamic Marketing, 1*, 9–24.

Stendhal: O Vermelho e o Negro. Lisboa. (1995).

Stephenson, M. L., & Ali-Knight, J. (2010). Dubai's tourism industry and its societal impact: Social implications and sustainable challenges. *Journal of Tourism and Cultural Change*, *8*, 278–292. https://doi.org/10.1080/14766825.2010.521248

The Emirates Group: *The Emirates Group.* https://www.theemiratesgroup.com/

Theobald, W. F. (Ed.). (2013), *Global tourism.* Butterworth Heinemann.

UNESCO: UNESCO and the international year of sustainable tourism. https://en.unesco.org/iyst4d

United Nations Environment Programme (UNEP), & World Tourism Organization (UNWTO). (2005). Making tourism more sustainable A guide for policy makers.

Vij, M., Upadhya, A., Vij, A., & Kumar, M. (2019). Exploring residents' perceptions of mega event-Dubai Expo 2020: A Pre-event perspective. *Sustainability, 11.*

Weatherby, C., Eyler, B., & Burchill, R. (2023). *Exporting renewable energy to the global south.*

Weaver, D. (2007). *Sustainable tourism.* Butterworth Heinemann.

The World Economic Forum: The Travel & Tourism Competitiveness Report. (2019).

World population review: World population by country (2024). https://worldpopulationreview.com/

WTO. The concept of sustainable tourism. http://www.world-tourism.org/sustainable/concepts.htm

Zaidan, E., & Development, P. (2016). *The impact of cultural distance on local residents perception of tourism development: The case of Dubai in UAE.*

Chapter 2

Ecotourism in Zimbabwe: The Challenges of Balancing Social, Economic and Environmental Goals

Providence Warinda[a,1], *Livingstone Mhonda*[a,2], *Tinashe A Kodzanai*[b,3], *Tarisai Gundiwa*[b,4] *and Anita Rinomhota*[a,5]

[a]University of Zimbabwe, Zimbabwe
[b]Harare Polytechnic College, Zimbabwe

Abstract

Over the past two decades, tourism in Zimbabwe was affected by the political and economic situation resulting from bad publicity. As a result, most ecotourism establishments under the CAMPFIRE program collapsed. Of the ecotourism establishments that survived, the revenues are decreasing such that incomes accruing to the communities have become insufficiently low. Hence, it became difficult to persuade the individuals to work collectively towards conserving and preserving the natural heritage. Ngomakurira Hills (a heritage site) was used as a case study for this particular chapter. The study used a qualitative methodology including observation, and literature review to enable the researcher to obtain the required information. Results of the study reveal that the community feels there has been a divergence from the original goals of the project which were community involvement, employment creation, improvement in income and standards of living. None of these goals were achieved, most of the respondents highlighted there has not been a change brought about by the project. The major factors affecting the potential for sustainable ecotourism at Ngomakurira were political and

[1]https://orcid.org/0009-0002-6837-313X
[2]https://orcid.org/0009-0005-4350-6348
[3]https://orcid.org/0009-0000-9287-3324
[4]https://orcid.org/0009-0005-0926-6838
[5]https://orcid.org/0009-0005-6172-0954

Sustainable Tourism, Part B, 19–43

doi:10.1108/978-1-83608-326-920241002

economic instability, lack of effective marketing and promotion, limited financial and human resources, corruption, lack of transparency, insufficient collaboration between the community, private and public sectors. The study recommends the development of an ecotourism model meant to ensure equitable distribution of ecotourism benefits, sustainable community support for the project by enhancing trans-generational preservation of natural and cultural resources.

Keywords: Ecotourism; heritage sites; community participation; conservation; sustainability; ecotourism model

1. Introduction

Ecotourism is a form of sustainable tourism that emphasizes the participation and the improvement of the quality of life of local communities in order to promote sustainable utilization of natural resources (Shereni & Saarine, 2021). In this respect, ecotourism represents incentive-based conservation thinking (Swatuk, 2005). Incentive-based conservation yields outcomes of employment, ecological services, compensation payments, health care, education, agroforestry and tourism development/promotion subsequently providing motivation to obtain for conservation (Ziegler et al., 2020). This is based on the belief that benefits derived directly from natural resources will encourage conservation and sustainable use of those resources (Stronza, 2007). Sustainable tourism differs from the conventional tourism in that it seeks to address the needs of tourists and the needs of the host communities. Host communities are the stewards of the natural resources and culture therefore they have to be prioritised in tourism developments.

Allendorf et al. (2012), Dimitriakopous et al. (2010), Mutanga et al. (2016) and Abukari and Mwalyosi (2018) brought out that the perceptions and attitudes of local people toward protected areas and conservation are modelled by several factors which include the local demographic characteristics the socio-economic characteristics, livelihood strategies. Perceptions and attitudes are also influenced by the local's experience of losses, perceived benefits of protected areas, perception of protected area existence, the perceived relationship of locals with protected area management and protected areas staff; the involvement of locals in protected area conservation, and the locals' level of awareness of protected areas; and ecological factors of distance from the park boundary.

Zimbabwe, renowned for its stunning landscapes, diverse wildlife and rich cultural heritage, has long been an attractive destination for tourists seeking unique and authentic experiences. With an array of national parks, including Hwange, Victoria Falls and Mana Pools, Zimbabwe offers a wealth of natural beauty and biodiversity (Zvinonyaukwa et al., 2023). Historically, the country's tourism industry has been centred on these natural assets, drawing visitors from all over the world. However, in recent years, a paradigm shift has been witnessed as Zimbabwe has increasingly embraced ecotourism, recognizing its potential not only for environmental conservation but also for entrepreneurial empowerment.

Tapfuma et al. (2023) brought out that ecotourism is gaining popularity in Zimbabwe because it places a strong emphasis on preserving the environment, supporting local communities and providing authentic experiences for tourists.

Zvinonyaukwa et al. (2023) supports the observation as he highlights that tourism development in Zimbabwe prioritizes environmental conservation and Community Development, the principles upon which the integration of ecotourism practices is based. Research in the areas neighbouring Matusadonha National Park demonstrates the contributions of ecotourism to economic growth and rural development. The results of the study underscore the capacity of ecotourism to uplift local communities while protecting natural resources (Zvinonyaukwa et al. 2023).

Ecotourism in Zimbabwe is uniquely positioned to leverage the natural and cultural assets of the country, offering innovative and sustainable experiences to tourists (Tapfuma et al., 2023). In addition, ecotourism provides ample opportunities for entrepreneurship, which is very important, given the need for economic diversification and poverty alleviation in Zimbabwe (Zvavahera & Chigora, 2023). The cultural and natural assets of the country can be harnessed by ecotourism to build a unique positive destination brand image which is essential for drawing tourists and investors alike (Chigora et al., 2023).

As Zimbabwean tourism continues to thrive, it has been observed that some forms of sustainable tourism such as ecotourism and community-based tourism are still lagging behind in terms of arrivals and proceeds from tourism. According to a report by Government of Zimbabwe (2017), the ecotourism projects under the CAMPFIRE program that were previously renowned as success stories are currently experiencing low revenues such that very few or no individuals in the community get any earnings from the establishments. Subsequently, there have been little change in existing local land and resource-use practices, as the ecotourism projects are providing only a modest supplement to local livelihoods, with the majority of the local community remaining dependent on external support for long periods, if not eternally (Mutanga et al., 2022). This results in reduced welfare of involved communities, especially vulnerable subgroups derailing the intended development and poverty alleviation gains (World Bank, 2017; Zimbabwe National Statistics Agency, 2016). Tchakatumba et al. (2019) indicated that the sustainability of ecotourism projects may be threatened by political crisis and economic disruptions. Tchakatumba et al. (2019) also alluded that the pulling out of donor funding and the land reform program adversely affected ecotourism projects. To ensure the success of ecotourism, host communities should also be empowered with knowledge, skills and power to make decisions.

A good ecotourism implementation framework designed to be a solution to biodiversity conservation and preservation of culture is therefore necessary (Chiutsi et al., 2011; Pasape et al., 2015; Puri et al., 2019). This study therefore sought to: (i) analyse how ecotourism is being implemented in Zimbabwe, (ii) identify emerging competing interests in the use of ecotourism/heritage site, (iii) explore the attitudes and perceptions of the Chinamhora community towards ecotourism/heritage site, (iv) investigate emerging practices that have negative impacts on the sustainability of ecotourism sites and (v) recommend an ecotourism implementation model meant to ensure sustainable community support for the project by enhancing trans-generational preservation of natural and cultural resources, and equitable distribution of ecotourism benefits.

In light of the aforestated considerations, this chapter seeks to provide a comprehensive model of the role of ecotourism in promoting heritage and cultural tourism in Zimbabwe using Ngomakurira Domboshava as a case study. This study aims to contribute valuable insights to the broader discourse on ecotourism in developing economies by investigating the dynamics, challenges and opportunities at play in this specific context. Additionally, it seeks to inform policy and decision-makers about strategies that can harness the potential of ecotourism to drive sustainable economic development, environmental conservation and community empowerment in Zimbabwe.

2. Literature Review

2.1 Ecotourism in Zimbabwe/Implementation of Tourism in Zimbabwe

Indigenous natural and heritage resources management in Zimbabwe have been characterized by an imbalance in the management and exclusion of local people and local leadership (Chiutsi et al., 2011). Community, involvement, unity and benefits sharing among the local people and other stakeholders are very important to ensure the success of ecotourism projects' goals (Mazambani & Dembetembe, 2010). Several definitions of ecotourism have been put forward. According to Sirakaya et al. (1999, p. 171), ecotourism is a new form of non-consumptive educational, and romantic tourism to relatively undisturbed and under-visited areas of immense natural beauty and cultural and historical importance for the purpose of understanding and appreciating the natural and sociocultural-history. Honey (2008) simply defined ecotourism as responsible travel to natural areas that conserves the environment and improves the well-being of local people. Weaver (2001) defined ecotourism as focusing on the learning experience, appreciation of the natural environment, environmental and social-cultural sustainable and viability of the operation. These definitions indicate that ecotourism projects are to achieve balanced socio-economic and environmental goals for the community in which it operates, and should also foster learning. Local people play a crucial role in fostering learning in ecotourism in cultural and heritage sites for they have indigenous knowledge. These pillars enhance the viability of ecotourism communities during the course of development however they may complement or compete with one another. To accomplish sustainability, a balance must be maintained between the three main 'pillars' of sustainable development (Hall et al., 2015).

This research study defines ecotourism as 'environmentally sustainable tourist activities in relatively undisturbed natural areas to enhance the appreciation of and learning about natural ecosystems while providing benefits to local communities and promoting businesses'.

2.2 Economic Goals of Ecotourism

The major economic goals of ecotourism include local economic growth, development, productive business and employment opportunities especially for the locals (Stronza & Gordillo, 2008; Weaver, 2014). Once the benefits are distributed equitably within the host communities (Das & Chatterjee, 2015; Honey, 2008), locals will become important partners in the protection of both the natural and cultural environments that form the basis to the community-based ecotourism projects (Chiutsi et al., 2011). However, the realisation of economic benefits from community-based ecotourism may be greatly reduced by a variety of factors including limited access of the poor to the tourism market (Boley & Green, 2016), lack of commercial viability for their product in terms of value and price (Masud et al., 2017), weak marketing capability (Mwaipopo, 2016), lack of intergovernmental suitable policy framework and inadequate knowledge about tourism and service skill, managing and implementing at local level (Anderson et al., 2002; Autthapon & Suthida, 2010). Weaver (2014) and Tietenberg and Lewis (2018) highlight that communities are usually allocated degraded resources that have no other claimant, while 'valuable natural resources create incentives for central actors to retain control over them', Literature also indicates that there are low-level economic benefits if any, from ecotourism (Weaver, 2014; Wondirad, 2017). According to Wondirad (2017), the overall combination of benefits has to be more than the costs incurred by communities, and that cost and benefits must be shared equitably between the stakeholders (Hewlett & Nicholls, 2011; Shackleton et al., 2002). Mbaiwa et al. (2008) states that unfair benefit distribution damages social capital and endangers the sustainability of ecotourism projects partly because many people may opt out of the project or become free riders.

2.3 Social Goals of Ecotourism

Equity, empowerment, accessibility, participation, sharing cultural identity, institutional stability and local community involvement are the major social goals of ecotourism (Mbaiwa et al., 2008; Sharpley, 2015). Ecotourism also seeks to create a relationship between locals and tourists through interaction with local people playing an active role in educating tourists about their culture. Community-Based Ecotourism (CBET) is an ecotourism subtype that places an additional emphasis on the local communities, tourists and other stakeholders' involvement in all aspects of an ecotourism project (Buckley and Weaver, 2003; Kiss, 2004; Muzvidziwa, 2013; Sharpley, 2015). Honey (2008) posits that ecotourism may contribute to the psychosomatic emancipation of local people by improving their sense of self-esteem, confidence and pride for their cultural and natural heritage, strengthening of social capital within the village and Weaver (2014) political empowerment.

Despite the social benefits that ensue, it has to be borne in mind that ecotourism is a commercial activity and that indigenous cultures are part of the attraction, whose cultures maybe be commoditised in the process of producing an experience for the tourists. Critics of the culture commercialisation argue that the hosts' culture will be eroded as it becomes an economic activity devoid of deeper meaning (Hanifah &

Webb, 2017). King and Stewart (1996) concur that the commodification of culture implies a change in the meaning of their environment and a shift in the relationship between the indigenous people, their institutions and their environment, from one working with the land to one working for tourists (Cousins et al., 2009).

2.4 Environmental Goals

In the opinion of Weaver (2014), ecotourism contributes to the conservation of the ecosystem while respecting the integrity of host communities. This means that 'it should contribute vigorously to the preservation and conservation of natural and cultural heritage' (MacLaren, 2002). Literature has it that ecotourism is both a conservation strategy as well as a developmental tool (Muzvidziwa, 2013). According to (Miller, 2012; Porter & Salazar, 2005) ecotourism minimises the environmental impact of tourism and contributes to environmental and cultural awareness among local residents and visitors.

Ecotourism may also have negative impacts such as pollution, soil erosion, loss of natural habitat, increased pressure on endangered species and heightened vulnerability to forest fires if not properly monitored (Khanal & Babar, 2007). The adverse impacts from tourism happen when the level of visitor use is larger than the environment's ability to cope within the acceptable limits of change (carrying capacity) (Hill & Gale, 2016). The critical issue is that the active involvement and equitable benefit of the local communities in ecotourism projects should involve deliberate steps to empower them so as to minimize impacts, through the choice of activity, equipment, location and timing; group size; education and training; and operational environmental management (Fennell, 2014).

2.5 Threats to Sustainable Tourism in Zimbabwe

Growth and expansion of cities in developing countries are threatening the sustainability of tourism activities due to loss of animal natural habitat and biodiversity (Okhiria et al., 2016). Rapid growth of tourism industry in developing countries is a major source of environmental pressures especially on natural resources (Mondal, 2018) because most of the developments are happening in ecologically fragile areas. Various factors have threatened sustainability of ecotourism in Zimbabwe and these include the Coronavirus (COVID-19) (Mudzinge et al., 2021), infrastructure development, climate change and behavioural change. Conservation of wildlife induces costs on local communities (livestock depredation, killing of people by wild animals and restriction to natural resources) creating hostility and antagonistic behaviour towards sustainability use of natural resources (Mutanga et al., 2022). In the research done by Mutanga et al. (2022), illegal hunting, lack of local people participation, lack of community benefits and destruction of animal habitats came out as champion threats to sustainable wildlife tourism. Tourism development in Zimbabwe, across all sectors of tourism, is stamping permanent marks that cannot be erased, depriving future generations their ability to benefit economically, socially and environmentally from natural resources and all gifts of Mother Nature (Kabote et al., 2016). Sustainable tourism development can be promoted with the support of local communities (Machnik, 2021) as they are the chief custodians of the natural resources within their locality.

2.6 Theoretical Framework

Gregor (2002) refers to underpinning theories as theories to comprehend the social setting in qualitative research projects. The theories aim to explain 'how' and 'why' things happen the way they do. The theory was considered as a lens to help with data interpretation and analysis in research. These theories aid in excavating the dependencies and linkages that exist among actors in a given setting. Their capacity to close the gap between assumptions and action prediction within a context is critical. The research, based on the aforementioned idea of underpinning theory, identified the following theories that can aid in the research process. These include knowledge-based theory, sustainable development theory and empowerment model.

2.7 Knowledge-Based Theory

The knowledge-based approach theoretical framework is an approach that analyses ecotourism from an objective and holistic approach (Chiutsi et al., 2011). This framework singles out the strengths and weaknesses of all types of tourism. It then uses scientific knowledge to identify the best tourism modes for each destination. Further, knowledge-based approach theoretical framework uses the main features of ecotourism with the sustainable development paradigm to create a measured growth approach for applying concepts to any destination. Alternatively, knowledge-based approach theoretical framework identifies and evaluates the destination's environment and sociocultural capacity. The approach considers tourism as a complex social phenomenon where knowledge is the basis for sustainable destination development (Jafari, 2001). This school of thought posits that through the generation and intelligent application of knowledge (on customer needs, collaborating suppliers, environmental and human and cultural resources) information asymmetries between stakeholders can be reduced.

2.7.1 Sustainable Development

Sustainable development is also part of the framework as it is premised on the triple bottom line model which is financial, social and environmental performance. Ecotourism supports the idea that the destination's financial viability must be reached without sacrificing the natural and sociocultural environments (Ali, 2009; Wall, 1997). Ecotourism is a form of tourism in which people visit undisturbed and untouched natural areas. Essentially, it means ecotourism focuses on socially responsible tourism to minimize the hazards and impacts of commercial mass tourism. However, the main aim of ecotourism is environmental sustainability through the management of resources with the assistance of local people, tourism operators and even policymakers. Ecotourism is premised on three important aspects (Shu, 2002). The first aim of ecotourism explain that various ecosystems existing in different destinations are important and the need to preserve cultural and biological diversity of the destination. The second aim of ecotourism is to improve the productivity of an ecosystem based on the notion that person and species have an important role to play in environmental sustainability. According to Riaz (2022) this is known as the concept of biodiversity whose aim is to promote the sustainable use of biodiversity by creating jobs and providing business opportunities. This will subsequently contribute to the greater good of all forms of life in an environment. The third aim of ecotourism is ensuring that benefits from ecotourism are provide to and shared with the local people as a way to ensure sustainable ecotourism development. The emphasis of ecotourism here is to ensure that the locals have plenty of opportunities to be involved in the planning and management of ecotourism benefits (Shu, 2002). As a result, the sustainable development model can be used to understand the evolution of the concept of ecotourism. Further, the model helps to identify and distinguish between various definitions of ecotourism and the challenges or criticism associated with such definitions. Therefore, framework is useful when we want to create an objective and holistic definition of ecotourism that could be essential for stakeholders in any destination.

2.7.2 Empowerment Framework

Community empowerment is important for ecotourism to achieve sustainable tourism development (Boley et al., 2015; Li & Hunter, 2015). Empowerment can be described as the means by which an individual or a community's capacity to make choices and transform those choices into desired actions leading to desired outcomes is enhanced (Rachmawati, 2020). Empowerment can be defined as process meant to improve community capacity and ability to participate in tourism development (Boley & McGehee, 2014; Timothy, 2006). Empowerment however can also be described as an outcome where it means the impact that an empowerment activity does to the community (Alsop & Heinsohn, 2005; Scheyvens, 2002). A review of literature done by Aghazamani and Hunt (2017) empowerment is evaluated mainly in terms of outcomes rather than process. Scheyvens (1999)'s empowerment framework is one of the most commonly used framework to measure the empowerment outcomes.

The framework was developed to provide an instrument to define the effectiveness of tourism initiatives in terms of their impacts on local communities (Scheyvens, 1999). The framework provides an instrument to define the effectiveness of tourism initiatives in terms of their impacts on local communities (Scheyvens, 1999). According to Scheyvens (1999)'s empowerment framework model, there are economic, psychological, social and political aspects of any industry (Fennell, 2014). Economic empowerment measures the extent to which ecotourism is able to provide consistent and equitable economic benefits for local communities. Psychological empowerment describes how ecotourism is providing optimism and hope for residents living in any community. The community feels that its traditions and culture must be respected and appreciated by visitors (Fennell, 2014; Scheyvens, 1999). Such positive attitudes encourage collaboration and cooperation that can be used for the growth and benefit of any destination. Social empowerment refers to the capacity of tourism to enhance community interactions and/or collaboration. As a result, ecotourism funds must concentrate on the development projects that bring about community harmony, community cohesion, well-being, resilience, social capital and solidarity (Boley & McGehee, 2014; Scheyvens, 1999). Lastly, there is a political empowerment which refers to the community's ability to control the direction, type and level of tourism development in their area (Boley et al., 2015; Scheyvens, 1999).

3. Methodology

In order to provide answers to our quest for improving implementation of good ecotourism practices around protected areas, we found it necessary to use field research, in-depth interviews, direct observations, and a focus group discussion with the village members. To quench the zeal for unearthing rich data from the community, exploratory research design was used (Bryman & Bell, 2011). Secondary data was also used to test theories (Cresswell & Cresswell, 2018). The nature of the research prompted the researchers to use qualitative research methodology. In order to identify the emerging competing interests at Ngomakurira, literature review was done in order to understand how other scholars investigated competing interests. Observations were used because the observer can freely observe things and respondents in the natural setting (Robson, 2002). More so, observation takes place while things are happening, allowing the researcher to study phenomena in a naturalistic way (Blanche et al., 2006). Observations allowed the researchers to get first-hand accounts of events and gaining insight into verbal and non-verbal communication and events (Bryman, 2016).

3.1 Study Area

The research was confined to communities around Ngomakurira (drums sounding) hills. The Hills are located approximately 30 km Northeast of Harare in Chinhamora village which is in Goromonzi District of the Mashonaland East Province. Ngomakurira Hill was chosen mainly because its national heritage site where ecotourism is being promoted. Ngomakurira Hill is one of the ecotourism projects that received support from USAID through the CAMPFIRE Development Fund from 2000 to 2003 and the implementers of the project were the Rural District Council (RDC) and Community Trust (Mazambani & Dembetembe, 2010). Previously, Ngomakurira Hill was used for grazing, fuelwood collection, harvesting of wild fruits and uncontrolled visits for recreational and religious purposes. Ngomakurira Hill was then converted into a properly managed tourist facility. It was envisaged that by controlling visits to the area, deforestation of the hills and disturbance of ancestral burial grounds would also be reduced resulting in preservation of the site as a tourist attraction.

3.2 Sample Size and Sampling Procedure

The population consisted people of Chinhamora Community aged 18 years and above because the research required selection of the most appropriate people as targeted sample to obtain most effective answers (Slade & Sergent, 2018). The researchers randomly selected and interviewed 55 male and females. The Chief, Headman and village elder in their capacity as leaders and custodians of the heritage site were also interviewed.

3.3 Data Collection Methods

Data was collected using both primary and secondary methods. Interviews and observations were used to gather primary data. Interview guides were developed and used to foster correctness and effective data gathering (Barnham, 2015). Interviews were audio-recorded with the consent of participants. Part of the data was also collected through review of related literature, and textbooks, journals, articles and book chapters were used for this purpose. Unstructured observation complemented the data gathered from interviews.

3.4 Data Analysis and Presentation

Qualitative techniques were employed for data analysis. Themes emerged through data reduction and this was made possible through continuous data collection and analysis. Audio files were downloaded then transcribed verbatim into English. A pilot study was conducted to help ensure that the interviews would collect the required data as accurately as possible. The researchers adhered to ethical practices by structuring the interview guides in such a way that the respondents participated voluntarily and anonymously. Relevant authorities were contacted before entry into the field. Research consent was directly negotiated with the traditional leadership of the Chinhamora Village; the Chief gave the researchers the go ahead to conduct research. Collected data was triangulated using validation strategy to help build coherent justification to themes. During coding, respondents were coded using pseudonyms. As a means of increasing credibility and validity, the coding process has been clearly explained, showing how themes and codes were generated (Mortensen, 2021). Data was analysed thematically using manual methods according to Braun and Clarke (2006)'s six phases of thematic analysis. Since the research was driven by research questions, the top-down approach to data analysis was deemed most effective. Various themes emerged in relation to social, economic and environmental impacts of ecotourism to local communities.

4. Presentation of Findings

4.1 Main Economic Activities of in Chinamhora

Results indicate that the most common economic activities for locals living around Ngomakurira Hill include agriculture (crop farming and horticulture), craft work, buying and selling, carpentry, building and welding. The respondents highlighted that very few locals are formally employed while there is quite a number of self-employed and unemployed locals in the area. Results also indicate that there are respondents who have no economic activity at all and can be potential sources of labour. We observed that there are some farmers who sell their produce to tourists who visit the Ngomakurira hills. The Chief indicated that their community has people with a lot of talents which can be of help to the development of the area. Some of the community members were involved in the construction of a Gazebo which later collapsed because of lack of maintenance. One of the key informants was an elderly village member who is no longer involved in any economic activities, but she however has valuable and interesting information on the history of Ngomakurira Hill which has not yet been documented but could add value to the experience of tourists.

4.2 Uses of Ngomakurira

Ngomakurira hills is used by different users for different purposes, it is also considered to be a sacred mountain. Key informants revealed that before the ecotourism project, Ngomakurira hill was used by locals for religious purposes, as a source of firewood, source of wild fruits and vegetables. Currently the mountain is used mainly for recreation and religious purposes. The Chief narrated how important Ngomakurira hill is for the local people of Chinamhora and other visitors;

> Ngomakurira Hill is well known for its spiritual powers and people from Chinhamora do the religious festivals in the mountain since it is their ancestral burial place. The mountain is not only used by the people of the Traditional Religion only but even Christians use this mountain, they usually come for all-night prayers at the mountain. We have Christians from the Apostolic sect (Mapostori), and Pentecostals who usually visit the mountain for spiritual purposes. Some people are prescribed by our traditional healers to go to the mountain to do some rituals or washing themselves in the river that flows on top of the mountain. The mountain is also used for recreation purposes, one can burn a lot of calories climbing this mountain.

4.3 Perceptions of the Compliance With Rules and Regulations on the Use of Ngomakurira Hills

There are rules that are supposed to be observed when visitors are at Ngomakurira Hills. In honour of its cultural significance, users of this hill should treat it with respect. People are not allowed to litter, light fires, deface rock paintings and cut down trees. The key informants had different views on compliance. The tour guide indicated that most tourists who come for recreation purposes comply with most of the rules of the mountain. The tour guide observed that religious groups often do not comply with some of the rules, they litter on the mountain as well as lighting fires during the night for warms when they do their prayers or rituals. We observed some littering in the river on top of the mountain where people come to throw different kinds of items such as rice, ritual cloths, as a spiritual act called '*kurasirira*'. The tour guide indicated that from a religious perspective there is nothing wrong with this kind of littering however from an environmental perspective it is pollution, and from a business perspective this kind of pollution may affect the appearance of the area and may even scare away tourists. Observations made on the site revealed that, there is not enough written down information on the rules and regulations that should be followed on the site. Tourists are told the Do's and Don'ts in the mountain by word of mouth and only two visible signage that advise visitors to treat the area with respect and not to litter. The Headman also highlighted that there are some locals who cut down trees on the mountain. He also indicated that there are some visitors defacing the rock paintings who believes its being done by people from neighbouring villages. The head man also thinks that there are some tourists who just throw litter on the mountain for no reason.

The villagers believe that religious visitors, village neighbours and tourists are the violators of the rules and regulations that have been put in place to ensure conservation and preservation of the heritage site. Religious visitors were perceived to be the ones who are involved in lighting fires in caves resulting in the defacing of rock paintings. This is because they camp at night and need to light a fire for warmth and protection from dangerous insects and animals. Key informants believe that religious groups from the apostolic sect and African traditional religion, and people from neighbouring villages are the common violators of rules and regulations. The Mapostori come and have their all-night prayers in the caves where they light fires while people who come for their rituals leave things such as red clothes, black goats and hens.

5. Management and Participation of Locals

Our results from key informant revealed that Ngomakurira hill is being managed by the National Museums and Monuments. The Chief and village members indicated that the community is not involved in the management of Ngomakurira Hills at all. The only time the community was involved is during the initial stages of the CAMPFIRE program were the community built a gazebo which was never used after it was built. Representatives from the village indicated that during the inception of the project community members were involved in the maintenance of the site, cutting grass, picking up litter and guarding the site. There was a committee that was responsible for organising the community's participation. However, the committees dissolved when there was nothing tangible that was coming from the project. This demotivated the community since they were never paid or received any benefits from the project. The local CAMPFIRE committee that was established has never been functional. To date Ngomakurira is still open for tourists however the community is not getting any proceeds from the site. The proceeds from entry fees goes to the National Museums and Monuments.

6. Perceived Economic Impacts Versus Actual Impacts

When the ecotourism project started local people of Chinhamora expected that the ecotourism project would bring employment opportunities, business opportunities, diversified livelihood strategies and improved standard of living. Respondents were asked to indicate whether the ecotourism project had any economic impacts. The village members expressed their dissatisfaction with the progress of the ecotourism project. The community had high hopes which never came to a reality. The local people expected that the project would bring direct and indirect employment, business opportunities, ultimately improving the standards of living. The village members living adjacent to the heritage site indicated that the only positive thing brought about by the project is the electricity grid which is now closer. However, this is only applicable to those living close to the heritage site. Locals from neighbouring villages are not benefiting anything. The village head expressed how he had hopes for job creation for locals:

> When Ngomakurira Ecotourism Project was initiated, we were very happy as a community and the project seemed to be very promising. Our members participated in the construction of a Gazebo which was going to be used by the locals to exhibit and sell their products. There were also plans to build lodges. Locals living adjacent to Ngomakurira hills people were also prepared to host tourists and providing accommodation, food and beverage services. However, the project collapsed before we could enjoy these benefits. The project had potential to directly employ a number of locals, while some locals were to benefit from service provision and selling their produce.

The villagers are however hopeful that if something is done the perceived benefits may become a reality. They pointed out that the heritage site has potential to attract more tourists. They also highlighted what they feel must be done to ensure the sustainability of the project. The villagers believe that if more security guards are employed on the site, this might protect the site from further defacing of rock paintings, vandalism, deforestation and pollution. They also think that if the place is well marketed more tourists can come to the area. The respondents also indicated that they are more than willing to participate in the development of the ecotourism project. They hope that developments such as lodges, campsites will be done on the site. The village head also highlighted that the monument is losing much revenue because National Museums' site offices close at 6 p.m. and cannot collect entrance fees from visitors who camp on the mountain during the night.

7. Social Impacts of Ecotourism

Results from community members indicated that ecotourism had brought a positive change in the reputation of the village since people are now aware of the area and visit the place for religious and recreation purposes. The elderly in the village indicated that they had no problems with tourists who came to visit the heritage site. The elderly in the village however expressed their concerns with outsiders who come for ritual purposes. They believe that some of the rituals done by outsiders bring bad luck into their communities. The Chief also indicated that tourism activities affected the sacredness of the site since they do not observe the presence of their ancestors when doing their tourism activities. He indicated that:

> The place has lost it power and sacredness. Long ago we used to hear drum beats on the mountain during a certain time of the year. The traditional leaders and village elders would perform rituals and a lot of things would happen on the mountain. Now this is no longer happening. The place is slowly losing its cultural value.

Some of the villagers however indicated that ecotourism had brought no change in local people's confidence, self-esteem and pride because locals do not interact with any tourists and are not being involved in the management of Ngomakurira. Ecotourism has brought a positive change in the reputation of the village however villagers believe that it has also attracted illegal activities such as theft. One of the village members revealed that the Gazebo they had built was vandalised. The chief also said that there was no collaboration between the local leadership and the National museums and Monuments. At one time the chief was proactive and contacted the National Museums and Monuments to resolve some of the security issues on the mountain. Unfortunately, the authorities did not respond.

8. Impacts of Ecotourism on the Environment

The focus group discussion with the local community revealed that ecotourism has had a negative impact on the preservation of the heritage site since visitors who come deface paintings, light fire and litter on the mountain. Some of the villagers assume that some tourists who come for religious and ritual purposes are involved in the cutting down of trees and lighting fires on the mountain. However, some villagers felt that ecotourism had brought about a change in the preservation of the heritage site since its now being guarded by police. The Chief and the village head believe that the present-day security methods are not enough to protect and preserve the heritage site. The site was more secure under the protection of their ancestors since the ancestors had their own way of dealing with the violators of the rules. On the other hand, they hoped that their local people would provide manpower for security on the most sacred places.

9. Discussion

The understanding of the residents' attitudes towards ecotourism is very important to ensure the sustainability of tourism. Knowledge, attitudes and perceptions of local residents towards the impacts of ecotourism are important in planning and policy consideration for successful development, marketing and operation of existing and future tourism program. Empirical findings have revealed that there is a huge gap in the implementation of ecotourism. Community benefits are an important factor in ecotourism and any tourism establishment which does not benefit the community seize to be a sustainable tourism project. The study has revealed that ecotourism is still imaginary and not a reality in Ngomakurira. The project has been labelled 'ecotourism' however in reality it is just conventional tourism, whereby communities do not participate in the management and provision of services to tourism activities. Tien et al. (2021) noted that although the concepts of sustainable tourism development have many advantages, the socio-economic, political and institutional environments that exist in underdeveloped nations make it extremely challenging to put these principles into practice.

The major goals of ecotourism are local economic growth development, productivity and employment opportunities (Fennell, 2014). Ngomakurira ecotourism project aimed at creating employment and generate income for the local community through tourism. Empirical evidence from this study shows that the project has failed to create employment, generate income for locals and has failed to improve the standard of living of the locals. Communities are failing to benefit from ecotourism which include political issues, low tourist returns resulting in low revenues and marketing and promotion challenges. Chiutsi and Mudzengi (2012) argued that most ecotourism establishments were not economically viable because of few tourists being attracted to these ecotourism sites. Warinda and van der Merwe (2023) noted that local communities have a lot that they can offer to enhance the tourism experiences; however, major challenges that communities face is packaging their expertise into an attractive tourism product. Creativity and innovativeness are crucial to ensure the competitiveness of a destination.

Reichel et al. (2008) state that the viability of ecotourism is ascribed to high levels of tourist satisfaction a large number of visitors to generate financial profit. Ngomakurira hill is a popular site for many Christians in the apostolic sect who use the mountain mainly for their all-night prayers who are not willing to spend and support local businesses. Domestic tourism in Zimbabwe is low and locals need to be encouraged to travel through intense marketing and promotion. Domestic tourists can make a huge contribution to the industry. The retraction of donor contribution in the project has created a marketing gap which the communities are unable to address due to their lack of understanding of the tourism industry in general and tourism marketing (Nyaruwata, 2011).

This study revealed a host of challenges experienced by the ecotourism project that included inflexible bureaucratic structures, lack of transparency and accountability in the management of funds and exclusion of the locals from the management of the site. Buckley (2013) stated that frequently benefit streams are dictated by the government that most often officially own the resource. There is little understanding by the host communities on the amounts that accrue to them, and they do not have the negotiating skills and knowledge to assure and secure a reasonable share of profits. Unequal benefit sharing among stakeholders is a major threat to sustainability in the development of community ecotourism projects. Das and Chatterjee (2015) explained this view by stating that imbalances in benefits sharing make it difficult for local people to buy into conservation. On the other hand, very few benefits, if any, will accrue to rural communities due to the high amount of leakage in payment of earnings, wages and profits remitted to workers.

Ecotourism puts emphasis on local communities being involved in the planning, development and operation of tourism projects and contributing to the well-being of the community. It intends to involve the community in the management, encourage interaction with tourists and devolution of power. The local community had hoped that they would be involved in the management of the site and also guide visitors as to how to behave at Ngomakurira Hill without disturbing their ancestors. When the local community perceive themselves to be playing a peripheral role in the ownership and management of tourist resources, they may end up engaging in activities that threaten biodiversity. The implementation of the program downplayed the importance of working with communities to better manage and sustainably use their natural resources and then arrogating community benefits and rights.

One of the lessons learnt from the COVID-19 pandemic is the importance of involving local community in projects of conservation and preservation of heritage as the survival of these pillars became highly dependent on the locals' behaviours during the pandemic (Fletcher et al., 2020). It is important to get community buy-in and participation because they are the owners of the resources and they have indigenous ways of protecting their heritage resources. Local communities still believe that the pre-colonial natural and heritage resource conservation which was based on traditional beliefs was more effective because non-compliance attracted direct punishment from the ancestors. This system has been replaced by law enforcement through fines approach which is believed to be very porous, subject to corruption hence ineffective and disadvantageous the local communities (Matseketsa et al., 2019).

Mandondo (2000) notes that non-governmental organizations (NGO) have facilitated the state in taking over traditional resources using the legal system and international donor support. The sustainability of ecotourism projects is hinged on policies and institutional arrangements that promote the agenda of local people benefiting from their own resources and participating in tourism. Initiatives at the local level are not supported by initiatives at the macro level. This is true when we take a look at the legislative provision of natural resource management in this country. The local people are not empowered by the laws of this country to manage the resources (Murombedzi, 1999). In CAMPFIRE projects local people do not have a legal mandate and fiscal autonomy in the management of resources.

Rationale choice can be used to explain the conservation attitudes, of the local people. Mostly, community members will pick the choice that maximizes payout while minimizing risks (Hechter, 2017). Individuals participate in activities that benefit them and will not participate in an activity that may be costly to them in terms of time and money. When benefits are few or when there are no benefits, people tend to be reluctant to preserve and conserve their natural resources because they have nothing to lose (Pretty, 2003). People from Chinhamora village pointed out that the place was in need of more security personnel so that illegal activities happening on the site could be reduced. People are willing to cooperate and provide manpower to police the site however National Museums and Monuments is not willing to pay these people a small amount of appreciation.

When authority and responsibility are separated, institutions usually under-perform. The communities who are the producers of the natural and heritage resources have no much say in the management system, and their role has been limited to that of just initiating project plans while the state and its agencies retain the responsibility of approval, implementing and fiscal control. Ecotourism establishments can be efficacious in goals of conservation, community involvement and development once there is transfer of power, to local communities to enable them to directly benefit from the sustainable management of these resources.

Heritage sites are shared by tourists and locals who have dissimilar inclinations, motivations and experiences (Porter & Salazar, 2005). The presence of the different types of users might result in conflicts between the users. Religious groups may want to use the mountain for prayers and rituals while tourists want recreation. These users end up encroaching in each other's territory resulting in dissatisfaction on the other users. Local people from Chinhamora community believe Ngomakurira is a sacred place, where chiefs used to live, burial place for their chiefs, the source of life, an abode for communication with the spirits and also a habitat for wildlife, including the animals that are sacred for the Shona. The village elders feel that the place has now lost its sacredness because of people who visit the hill and do not respect their ancestors. This has resulted in extreme degradation of cultural values and beliefs, sacred places and biodiversity, and local people believe that this has caused bad luck in their community. Zhang et al. (2014) noted that tourism activities may be an inducer of conflict because of its overriding aim of making profit. The contradictory values, beliefs and goals may affect the protection of heritage resources. Commercialization versus authenticity conflicts always affect cultural heritage sites. The situation is worsened when communities fail to get their intended benefits, while the authenticity of the culture is being diluted.

It can be noted that there is no effective communication between the management authorities and the visitors. There are not enough written messages to guide users as to how to behave on the mountain. There is over reliance on verbal communication, but it is not always effective because the guide may forget to share all the rules and regulations to all visitors. People have their own ways of interacting with nature, shaped by their beliefs and attitudes. It is their beliefs and their attitudes that determine their behaviour towards around them (Harrill, 2004). People may not be aware of the implications of their actions on the site, for example when people light fire at night on the mountain it's a way to keep themselves warm and also protection from dangerous animals and they do not see anything bad in that. Graffiti on rocks is a way some people register their visit to a place.

10. Framework to Promote Sustainable Ecotourism Development and Management of Heritage and Tourism Sites

Fig. 2.1 shows a framework that can be followed to promote sustainable management and use of tourism and heritage sites.

Proper management of tourism heritage sites can be achieved when all the stakeholders are effectively participating in processes involved. The community people, community leadership and the National Museums and Monuments have been identified as key stakeholders of Ngomakurira Hills. To promote sustainable ecotourism at the heritage site, it is recommended that the identified stakeholders

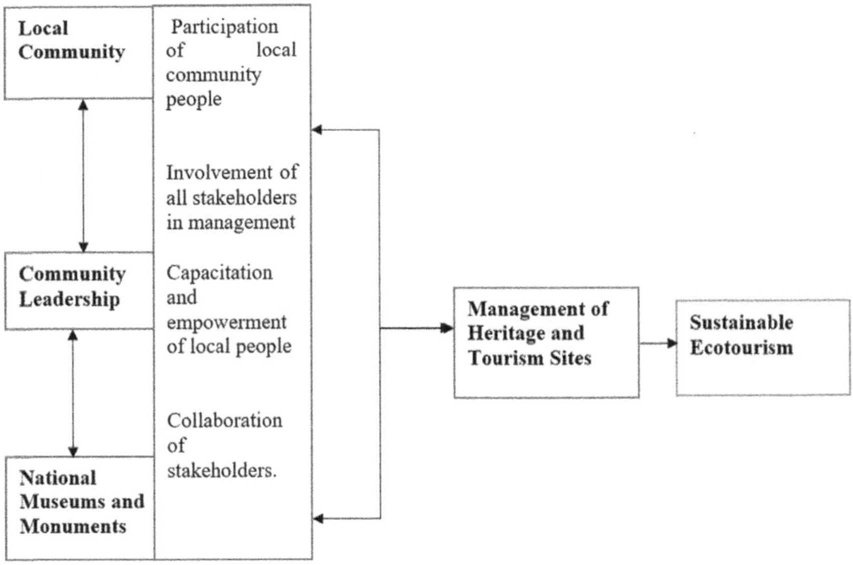

Fig. 2.1. Framework for Sustainable Ecotourism Development and Management. *Source:* Author's own compilation.

work together in the planning, management and operation of the Hills. Local people should be well capacitated and empowered with knowledge so that they may be able to take part in the tourism developments of Ngomakurira. If well capacitated, the local community can be able to manage and market their production through emerging information technology platforms. Local people must fully participate in the process to promote and instil the mentality of conservation to them. They have to feel that they are part of the process. Local people have proved to be the information hubs and have got the history of Ngomakurira which can enhance the tourism experience. Together with support from the National Museums and Monuments, the management of the heritage site will be sustainable. Greening the management and operations will also lead to sustainability of the heritage site.

11. Conclusion and Recommendations

Stronza et al. (2019) suggest that there should be a frame work for evaluation of ecotourism to ensure that it achieves its economic, social and environmental goals. They suggested that ecotourism needs to be clearly defined within the context of the socio-economic and demographic characteristics of the host community. A thorough research and assessment on the impacts of ecotourism development is essential so as to put in place strategies that maximize the positive impacts and minimize the negative impacts (Stronza et al., 2019). In the case of Ngomakurira hills, there is a need for incorporating the community, to empower the communities so that they become part of the tourism product.

An important lesson learnt from the Ngomakurira ecotourism venture is that local communities if left alone without technical support and business marketing capacitation from established tourism enterprises and conservation agencies will always find it difficult to maintain the overall sustainability of the project. In concurrence, Balint and Mashinya (2006), Mudzengi et al. (2021) note that local participatory decision-making and resource-governing institutions are fragile and therefore require continuous external support and capacity building from established conservation agencies in managing ecotourism initiatives. However, a sustainable self-funding as ecotourism venture is required in order to limit ecotourism ventures' dependence on external support.

Economically, it is expected that Ngomakurira ecotourism project should generate enough profits to be distributed in the community and facilitate overall development. However, this is not the case with the Ngomakurira ecotourism project because the returns are too low as there is low – market demand for this form of tourism (Tolkach et al., 2013). In agreement to this assertion, Tichaawa and Mhlanga (2015) comment that economic contributions from ecotourism ventures to the local communities are usually paltry because of weak provider – tourist business interactions. In light of this, we can only maximize the multiplier effect of ecotourism project while reducing economic leakages, through increasing business readiness among locals by strengthening the economic linkages in backward (suppliers-providers) and forward (providers-tourists) interactions (Holladay & Powell, 2013).

The National Museums and Monuments should work with the community and traditional leaders, and come up with detailed rules that must be followed when patronizing Ngomakurira hills. Further, there is need for concerted efforts of all stakeholders to come up with well-designed educational services targeting tourists (e.g. signage) and locals (e.g. workshops) because improved knowledge fosters supportive attitudes towards resource management, positive environmental behaviour and conservation philanthropy (Tichaawa & Mhlanga, 2015). For instance at the Ngomakurira ecotourism project there is need for adequate interpretation meant to shift attitudes from engaging in destructive activities like littering, deforestation and defacing of rock paintings.

Ngomakurira ecotourism project can improve on endeavours to attain sustainability if there is an established presence of strong networking platforms (Polnyotee & Thadaniti, 2015). These can take place within a community, between the community and supporting agencies, and among local CBT initiatives (Tolkach et al., 2013). Networking between the community and supporting public and private (e.g. tour operators, non-profit organizations) agencies is paramount for the long-term viability of the project (Holladay & Powell, 2013).

We cannot deny the fact that heritage sites are associated with some levels of spirituality, and therefore the role of traditional leaders must not be ignored. The study proposes a model which encapsulates the involvement, participation and collaboration of the local community, traditional leaders and the National Museums and Monument authorities in the effective management of heritage and tourism sites in order to attain sustainable ecotourism. Ngomakurira Hill is an ideal ecotourism destination due to its abundance of natural flora and fauna, and rich, diverse culture and heritage; conversely, ecotourism has enormous untapped economic, social and environmental potential. The study highlights the challenges posed by the lack of involvement, participation and empowerment of the local community and traditional leaders in the management of the ecotourism project. A framework was developed to ensure effective management links ecological preservation to financial gains. This not only empowers local communities but motivates local populations to preserve their resources.

References

Abukari, H., & Mwalyosi, R. B. (2018). Comaparing conservation attatudes of park-adjacent communities: The case of Mole National Park in Tanzania. *Tropical Conservation Science*, *11*, 1–14.

Aghazamani, Y., & Hunt, C.A. (2017). Empowerment in tourism: A review of peer-reviewed literature. *Tourism Review International*, *21*(4), 333–346.

Ali, H. M. (2009, March). Development of Arab water sustainability index using principal component analysis. In *Proceedings of the Thirteenth International Water Technology Conference IWTC13*.

Allendorf, T. D., Aung, M., & Songer, M. (2012). Using residents' perceptions to improve park-people relationships in Chatthin Wildlife Sanctuary, Myanmar. *Journal of Environmental Management*, *99*, 36–43.

Alsop, R., & Heinsohn, N. (2005). *Measuring empowerment in practice: Structuring analysis and framing indicators*. Available at SSRN 665062.

Autthapon, S., & Suthida, J. (2010). *Poverty alleviation through Community based ecotourism in the trans-boundary protected areas*. LAP Lambert Publishing.

Balint, P.J., & Mashinya, J. (2006). The decline of a model community-based conservation project: Governance, capacity, and devolution in Mahenye, Zimbabwe. *Geoforum*, *37*(5), 805–815.

Barnham, C. (2015). Quantitative and qualitative research: Perceptual foundations. *International Journal of Market Research*, *57*(6), 837–854.

Blanche, M., Blanche, M., Durrheim, K., & Painter, D. (2006). *Research in practice: Applied methods for the social sciences*. s.l.:Juta and Company Ltd.

Boley, B. B., & Green, G. T. (2016). Ecotourism and natural resource conservation: The 'potential' for a sustainable symbiotic relationship. *Journal of Ecotourism*, *15*(1), 36.

Boley, B. B., & McGehee, N. G. (2014). Measuring empowerment: Developing and validating the resident empowerment through tourism scale (RETS). *Tourism Management*, *45*, 85–94.

Boley, B. B., Maruyama, N., & Woosnam, K. M. (2015). Measuring empowerment in an eastern context: Findings from Japan. *Tourism Management*, *50*, 112–122.

Braun, V., & Clarke, V. (2006). Using thematic analysis in psychology. *Qualitative Research in Psychology*, *3*, 77–101.

Bryman, A., & Bell, E. (2011). *Business research methods* (3rd ed.). Oxford University Press.

Bryman, A. (2016). *Social research methods*. s.l.:Oxford University Press.

Buckley, C. P., & Weaver, D. B. (2003). *Nature-based tourism, environment, and land management*. CABI.

Buckley, R. (2013). Ecotourism and conservation. In *International Handbook on Ecotourism* (pp. 233–244). Edward Elgar Publishing.

Chigora, F., Nyagadza, B., Katsande, C., & Zvavahera, P. (2023). Building positive Zimbabwean tourism festival and event destination image and equity. In *Resilient and sustainable destinations after disaster: Challenges and strategies* (pp. 63–74). Emarald Publishing Limited.

Chiutsi, S., Mukoroverwa, M., Karigambe, P., & Mudzengi, B. K. (2011). The theory and practice of ecotourism in Southern Africa. *Journal of Hospitality Management and Tourism*, *2*(2), 14–21.

Chiutsi, S., & Mudzengi, B. K. (2012). Community tourism entrepreneurship for sustainable tourism management in Southern Africa: Lessons from Zimbabwe. *International Journal of Academic Research in Business and Social Sciences*, *2*(8).

Community Development. A Case of the CAMPFIRE Programme in Zimbabwe. In Iscontour 2022 tourism research perspectives: Proceedings of the international student conference in tourism research (p. 176). BoD–Books on Demand.

Cousins, J., Evans, J., & Sadler, J. (2009). Selling conservation? Scientific legitimacy and the commodification of conservation tourism. *Ecology and Society*, *14*(1).

Cresswell, J., & Cresswell, J. (2018). *Research design: Qualitative, quantitative and mixed methods approaches*. s.l.:Sage.

Das, M., & Chatterjee, B. (2015). Ecotourism: A panacea or a predicament? *Tourism Management Perspectives*, *14*, 3–16.

Dimitriakopous, P. G., et al (2010). Local attitudes on protected areas: Evidence from three Natura 2000 wetlands sites in Greece. *Journal of Environmental Management*, *91*(9), 1847–1854.

Fennell, D. A. (2014). *Ecotourism*. Routledge.

Fletcher, R., Buscher, B., Massarella, K., & Koot, S. (2020). 'Close the tap!': COVID-19 and the need for convivial conservation. *Journal of Australian Political Economy*, (85), 200–211.

Government of Zimbabwe. (2017). *Zimbabwe vulnerability assessment committee (ZimVAC) 2017 Rural Livelihoods Assessment*.

Gregor, S. (2002). A theory of theories in information systems. In *Information systems foundations: Building the theoretical base* (pp. 1–20). Australia University.

Hall, C. M., Gössling, S., & Scott, D. (Eds.). (2015). *The Routledge handbook of tourism and sustainability* (Vol. 922). Routledge.

Hanifah, N. A., & Webb, A. B. (2017). Ecotourism and biopiracy: A legal perspective on the sustainability of tourism destination in Malaysia. In *Balancing development and sustainability in tourism destinations* (pp. 9–19). Springer.

Harrill, R. (2004). Residents' attitudes toward tourism development: A literature review with implications for tourism planning. *Journal of Planning Literature*, *18*(3), 251–266.

Hechter, M. (2017). Towards a sociological rational choice theory. In *The social sciences and rationality* (pp. 35–52). Routledge.

Hewlett, T. R., & Nicholls, S. (2011). *Benefits of empty promises: Ecotourism and local communities*.

Hill, J., & Gale, T. (2016). Ecotourism and environmental sustainability: An introduction. In *Ecotourism and environmental sustainability* (pp. 21–34). Routledge.

Holladay, P.J., & Powell, R.B. (2013). Resident perceptions of social–ecological resilience and the sustainability of community-based tourism development in the Commonwealth of Dominica. *Journal of Sustainable Tourism*, *21*(8), 1188–1211.

Honey, M. (2008). Setting standards: Certification programmes for ecotourism and sustainable tourism. In *Ecotourism and Conservation in the Americas* (pp. 234–261). CABI.

Jafari, J. (2001). The scientification of tourism. In *Hosts and guests revisited: Tourism issues of the 21st century* (pp. 28–41).

Kabote, F., Chipungu, P., Mamimine, P. W., & Muranda, Z. (2016). Human activities and sustainable tourism development at cultural heritage sites: Focus on great Zimbabwe. *Revista de turism-studii si cercetari in turism*, (22).

Khanal, B. R., & Babar, J. T. (2007). Community based ecotourism for sustainable tourism development in the Mekong region. *Policy Brief*, *1*.

King, D. A., & Stewart, W. P. (1996). Ecotourism and commodification: Protecting people and places. *Biodiversity & Conservation*, *5*, 293–305.

Kiss, A. (2004). Is community-based ecotourism a good use of biodiversity conservation funds?. *Trends in ecology & evolution*, *19*(5), 232–237.

Li, Y., & Hunter, C. (2015). Community involvement for sustainable heritage tourism: A conceptual model. *Journal of Cultural Heritage Management and Sustainable Development*, *5*(3), 248–262.

Machnik, A. (2021). Ecotourism as a core of sustainability in tourism. In *Handbook of sustainable development and leisure services* (pp. 223–240).

MacLaren, F. T. (2002). The international year of ecotourism in review. *Journal of Sustainable Tourism*, *10*(5), 443–448.

Mandondo, A. (2000). *Situating Zimbabwe's natural resource governance systems in history* (No. CIFOR occasional paper no. 32, p. 20). Centre for International Forestry Research.

Masud, M. M., Aldakhil, A. M., Nassani, A. A., & Azam, M. N. (2017). Community-based ecotourism management for sustainable development of marine protected areas in Malaysia. *Ocean & Coastal Management*, 136, 104–112.

Matseketsa, G., Mukamuri, B. B., Muboko, N., & Gandiwa, E. (2019). An assessment of local people's support to private wildlife conservation: A case of Save Valley conservancy and fringe communities, Zimbabwe. *Scientifica*, *2019*(1), 2534614.

Mazambani, D., & Dembetembe, P. (2010). *Community based natural resources management: Stocktaking assessment: Zimbabwe profile*. USAID.

Mbaiwa, J. E., Thakadu, O. T., & Darkoh, M. B. K. (2008). Indigenous knowledge and ecotourism-based livelihoods in the Okavango Delta in Botswana. *Botswana Notes and Records*, 62–74.

Mondal, M. S. H. (2018). *Social, economic and environmental sustainability: The implications to develop sustainable tourism in Bangladesh*. Research Gate.

Mortensen, D. (2021). *How to do thematic analysis of user interviews*. s.l.:The Interaction Design Foundation.

Mudzengi, B. K., Gandiwa, E., Muboko, N., & Mutanga, C. N. (2021). Towards sustainable community conservation in tropical savanna ecosystems: A management framework for ecotourism ventures in a changing environment. *Environment, Development and Sustainability*, *23*, 3028–3047.

Murombedzi, J. C. (1999). Devolution and stewardship in Zimbabwe's CAMPFIRE programme. *Journal of International Development*, *11*(2), 287.

Mutanga, C. N., Mamimine, P. W., Kwanisai, G., & Sambaza, P. (2022). *Re-thinking ecotourism and community development: The case of Ngomakurira Hills Project, Zimbabwe*.

Mutanga, N. C., Muboko, N., Gandiwa, E., & Vengesai, S. (2016). Beyond a single perspective to conservation relationship: Exploring factors influencing protected area staff and local community relationship in Zimbabwe. *International Journal of Biodiversity Science Ecosystem Services and Management*, *12*(3), 212–226.

Muzvidziwa, V. N. (2013). Eco-tourism, conservancies and sustainable development: The case of Zimbabwe. *Journal of Human Ecology*, *43*(1), 41–50.

Mwaipopo, R. (2016). Significant social and economic aspects of biodiversity conservation. *Western Indian Ocean*, 153.

Nyaruwata, S. (2011). Tourism, biodiversity conservation and rural communities in Zimbabwe. *Journal of Sustainable Development in Africa*, *13*(8), 225–238.

Okhiria, A. O., Paul, P. W., Obadeyi, J. A., & Afuye, F. O. (2016). Banking sector and hospitality industry in ondo state: Issues and challenges. *American Journal of Tourism Mangement*, *5*(1), 19–28.

Pasape, L., Anderson, W., & Lindi, G. (2015). Assessment of indicators of sustainable ecotourism in Tanzania. *Anatolia*, *26*(1), 73–84.

Polnyotee, M., & Thadaniti, S. (2015). Community-based tourism: A strategy for sustainable tourism development of Patong Beach, Phuket Island, Thailand. *Asian Social Science*, *11*(27), 90.

Porter, B. W., & Salazar, N. B. (2005). Heritage tourism, conflict, and the public interest: An introduction. *International Journal of Heritage Studies*, *11*(5), 361–370.

Pretty, J. (2003). Social capital and the collective management of resources. *Science*, *302*(5652), 1912–1914.

Puri, M., Karanth, K. K., & Thapa, B. (2019). Trends and pathways for ecotourism research in India. *Journal of Ecotourism*, *18*(2), 122–141.

Rachmawati, E. (2020, July). Tourism and community empowerment at Gunung Leuser National Park, Indonesia. *IOP Conference Series: Earth and Environmental Science*, *528*(1), 012001. IOP Publishing.

Reichel, A., Uriely, N., & Shani, A. (2008). Ecotourism and simulated attractions: Tourists' attitudes towards integrated sites in a desert area. *Journal of Sustainable Tourism*, *16*(1), 23–41.

Riaz, A. R. (2022). Ecotourism evaluation of concepts and impact on Germany. *Academy of Entrepreneurship Journal*, *28*(S6), 1–16.

Robson, P. (2002). *The economics of international intergration*. s.l.:Routledge.

Shackleton, S., Campbell, B., Wollenberg, E., & Edmunds, D. (2002). Devolution and community-based natural resource management: Creating space for local people to participate and benefit. *Natural resource perspectives*, *76*(1), 1–6.

Scheyvens, R. (1999). Ecotourism and the empowerment of local communities. *Tourism management*, *20*(2), 245–249.

Scheyvens, R. (2002). *Tourism for development: Empowering communities*. Pearson Education.

Sharpley, R. (2015). Sustainability: A barrier to tourism development. In W. R. Sharpley, & D. J. Telfer (Ed.), *Tourism and development. Concept and issues* (2nd ed., pp. 428–452). Channel View Publications.

Shereni, N.C., & Saarinen, J. (2021). Community perceptions on the benefits and challenges of community-based natural resources management in Zimbabwe. *Development Southern Africa*, *38*(6), 879–895.

Shu, G. U. O. (2002). A study on broad conceptions of ecotourism. *Tourism Tribune*, *1*.

Sirakaya, E., Sasidharan, V., & Sönmez, S. (1999). Redefining ecotourism: The need for a supply-side view. *Journal of Travel Research*, *38*(2), 168–172.

Slade, S., & Sergent, S. (2018). *Interview techniques*. s.l.:StartPearls Publishing LLC.

Stronza, A. (2007). The economic promise of ecotourism for conservation. *Journal of Ecotourism*, *6*, 210–230.

Stronza, A., & Gordillo, J. (2008). Community views of ecotourism. *Annals of Tourism Research*, *35*(2), 448–468.

Stronza, A. L., Hunt, C. A., & Fitzgerald, L. A. (2019). Ecotourism for conservation? *Annual Review of Environment and Resources*, *44*, 229–253.

Swatuk, L. A. (2005). From "project" to "context": Community natural resource management in Botswana. *Global Environmental Politics*, *3*(5), 95–124.

Tapfuma, M. M., Musavengana, R., & Magwaza, R. (2023). The Role of creative tourism through arts and crafts in promoting inclusive tourism in Zimbabwe. *Journal of Tourism and Cultural Change*, 1–20.

Tchakatumba, P.K., Gandiwa, E., Mwakiwa, E., Clegg, B., & Nyasha, S. (2019). Does the CAMPFIRE programme ensure economic benefits from wildlife to households in Zimbabwe? *Ecosystems and People*, *15*(1), 119–135. https://doi.org/10.1080/26395916.2019.1599070

Tichaawa, T. M., & Mhlanga, O. S. W. A. L. D. (2015). Community perceptions of a community-based tourism project: A case study of the CAMPFIRE programme in Zimbabwe. *African Journal for Physical Health Education, Recreation and Dance*, *21*(Suppl. 2), 55–67.

Tien, N. H., Viet, P. Q., Duc, N. M., & Tam, V. T., (2021). Sustainability of tourism development in Vietnam's coastal provinces. *World Review of Entrepreneurship, Management and Sustainable Development*, *17*(5), pp.579–598.

Tietenberg, T., & Lewis, L. (2018). *Environmental and natural resource economics*. Routledge.

Timothy, D. J. (2006). Empowerment and stakeholder participation in tourism destination communities. In *Tourism, power and space* (pp. 213–230). Routledge.

Tolkach, D., King, B., & Pearlman, M. (2013). An attribute-based approach to classifying community-based tourism networks. *Tourism Planning & Development*, *10*(3), 319–337.

Wall, G. (1997). Is ecotourism sustainable? *Environmental Management*, *21*(4), 483–491.

Warinda, P., & van der Merwe, C. D. (2023). Tourism and recreation in protected areas: An exploration of Community Based Tourism and local participation in Zimbabwe. *Studia Periegetica*, *43*(3), 65–86.

Weaver, D. B. (2001). *Ecotourism in the context of other tourism types. The Encyclopedia of Ecotourism* (pp.73–83). CABI Digital Library.

Weaver, D. (2014). *Romancing the wild: Cultural dimensions of ecotourism*.

Wondirad, A. (2017). Who benefits from the ecotourism sector in Southern Ethiopia? *International Journal of Tourism Sciences*, *17*(4), 276–297.

World Bank. (2017). http://data.worldbank.org/indicator/NY.GDP.PCAP.CD/countries/BT-NP-8S?display=graph. Accessed on 7 May 2018.

Zhang, H., Fu, X., Cai, L. A., & Lu, L. (2014). Destination image and tourist loyalty: A meta-analysis. *Tourism management*, *40*, 213–223.

Ziegler, J., et al (2020). *Can ecotourism change community attitudes towards conservation?* Fauna & Flora International.

Zimbabwe National Statistic Agency. (2016, September 12). http://www.zimstat.co.zw/sites/default/files/img/Zimbabwe%20Food%20Poverty% 20Atlas2016_FINAL_A4_3.pdf. Accessed on 15 September 2018.

Zvavahera, P., & Chigora, F. (2023). Agritourism: A source for socio-economic transformation in developing countries. *Oeios*.

Zvinonyaukwa, J., Musengi, K., & Mudzengi, C. P. (2023). Assessing the contributions of ecotourism growth and rural development offered by wildlife resources to people living in communities around Matusadonha National Park. *Journal of Sustainable Business and Economics*, *6*(2), 12–24.

Chapter 3

Carrying Capacity Studies for Sustainable Tourism: A Science-Based Management Approach to Sustainable Manta Ray Tourism in Raja Ampat, West Papua – Indonesia

Maulita Sari Hani[a,1] *and Syafri Tuharea*[b,2]

[a]Manta Ray Conservation Working Group of Raja Ampat, Indonesia
[b]Regional Technical Implementation Unit (BLUD-UPTD) Marine Protected Area Management of Raja Ampat, Indonesia

Abstract

Raja Ampat, located in West Papua – Indonesia, is renowned as a favoured tourist destination for manta ray tourism. In the years following the pandemic, there has been a noticeable increase in the number of visitors. Despite the benefits from tourists' expenditures, however, managing the influx of tourists poses an ongoing challenge in promoting sustainable tourism. This chapter focuses on carrying capacity studies examining manta ray tourism's economic, social and physical aspects in the Dampier Strait – Raja Ampat. Data was collected through on-site observations, interviews and surveys. The study aims to contribute to efforts and actions to reduce adverse effects on manta ray populations and their environments while advancing conservation, educational initiatives and responsible tourism practices for manta ray sustainable tourism. In summary, recommendations for sustainable tourism practices are proposed, including collaboration between stakeholders to continue research and monitoring, enforce guidelines and support conservation initiatives.

[1]http://orcid.org/0009-0008-4045-525X
[2]http://orcid.org/0009-0007-3190-778X

Sustainable Tourism, Part B, 45–66

doi:10.1108/978-1-83608-326-920241003

Keywords: Sustainable tourism; manta ray, conservation; carrying capacity; marine protected area; management

1. Introduction

Raja Ampat encompasses over four million hectares of land and sea, with a rich biodiversity of more than 1,400 fish species and 537 coral species, representing 75% of all known coral species (Hani et al., 2019a, 2019b). Marine tourism is the primary industry in Raja Ampat, emphasising the crucial role of the local economy. To ensure the longevity of marine tourism in Raja Ampat, all nature-based tourism activities, including manta ray tourism, are conducted sustainably, incorporating educational components and overseen by local tourism stakeholders to minimise adverse impacts resulting from user activities.

According to Hani et al. (2019a, 2019b), manta rays, serving as a conservation icon for this region, remain abundant in Raja Ampat's waters despite severe depletion in their populations elsewhere in the region. They have become a significant attraction for marine wildlife enthusiasts and contribute to the local economy, particularly in the Dampier Straits (Fig. 3.1). The government has safeguarded manta rays since 2014 due to their endangered status. Nearly all waters in Raja Ampat are crucial habitats for manta rays and are protected as conservation areas. Pre-COVID-19, tourist visits to manta ray sites, especially in Dampier Straits, were notably high, exceeding 10,000 visitors annually. The rise of manta ray tourism in Raja has become a challenge resulting in over-tourism, where the manta ray tourism sites face an influx of visitors surpassing their capacity.

The study aims to maximise the benefits of manta ray tourism while minimising adverse impacts. Carrying capacity serves as the maximum intensity, striking a balance for the long-term sustainability of nature; therefore, the study serves as a crucial comprehensive socio-economic and physical instrumental in mitigating unintended adverse impacts. The study will contribute to efforts and actions to reduce adverse effects on manta ray populations and their environments while advancing conservation, educational initiatives and responsible tourism practices for manta ray sustainable tourism. Recommendations are proposed for sustainable tourism practices, including collaboration between stakeholders to continue research and monitoring, enforce guidelines and support conservation initiatives.

2. Literature Review

2.1 Manta Ray at a Glance

Deakos (2010) classifies manta rays under the taxonomic family Mobulidae, consisting of two genera, Mobula (with nine species) and Manta (with two species). This group is characterised by a single lobe on each side of the head, wing-like pectoral fins, a terminal mouth, and a tail without a sting (Notarbartolo-di-Sciara et al, 2006). The Manta genus includes Reef manta ray (*Mobula alfredi*) and Oceanic manta ray (*Mobula birostris*). Mantas are filter feeders, using their frontal lobes to direct water

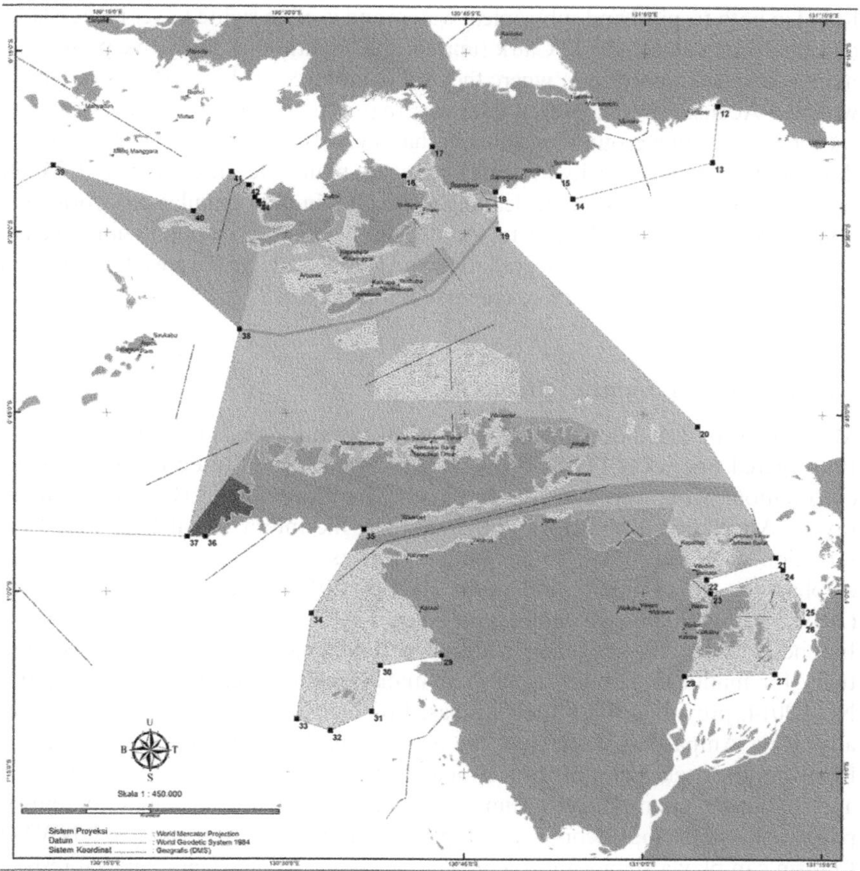

Fig. 3.1. Marine Protected Areas of Dampier Straits. *Source:* https://rajaampatmarinepark.com/dampier-strait-mpa/. Retrieved: 11 November 2023.

to their mouths, where planktonic organisms, such as copepods, mysid shrimps and arrow worms, are filtered. As per Dewar et al. (2008) and Deakos et al. (2011) identified that mantas have a lifespan of about 50–100 years, reaching sexual maturity between 10 and 25 years of age. Reef manta ray typically grows to an average wingspan of 3–4.5 m, weighing around 1.4 tons (Deakos, 2010), whereas Oceanic manta ray is larger, with a size of 7 m and a weight of 2 tons (Marshall et al., 2009). Oceanic manta ray is found globally in tropical and subtropical waters, making it a circumglobally species. The oceanic manta ray is predominantly pelagic and is observed in coastal and open waters, feeding in areas of high productivity (Dewar et al., 2008). These two species inhabit different environments, highlighting that Reef manta ray resides inshore around coral reefs, tropical island groups, atolls, bays and coastlines in tropical waters. On the other hand, the Oceanic manta ray is found in both tropical and subtropical waters and is often sighted over reefs, islands and the continental shelf. T. Clark (unpublished data) notes the active presence of mantas on cleaning stations, where they rid themselves of skin parasites or tend to wounds. Acoustic tracking data indicates that mantas migrate between cleaning stations and feeding grounds, residing in shallow reefs and surface waters during the day and migrating vertically to deeper waters at night (Dewar et al., 2008). Manta rays were facing various threats leading to declines in their populations worldwide. Some key factors contributing to the decline in manta ray populations include overfishing, bycatch, habitat degradation and global trade. In the past decades, there have been a number of efforts to conserve manta rays.

2.2 Conservation Status of Manta Ray

Manta rays are listed in Appendix II of the Convention on International Trade in Endangered Species (CITES, 2014), indicating controlled trade to prevent over-exploitation. Moreover, the Convention on Migratory Species (CMS) lists manta rays in Appendices I and II. At the same time, the International Union for Conservation of Nature (IUCN) designates them as 'Vulnerable' on the Red List, emphasising their conservation status (IUCN, 2011). According to Hani et al. (2019a, 2019b, 2020), the Ministry of Marine Affairs and Fisheries, Republic of Indonesia (MMAF), has officially declared Indonesia as the world's largest sanctuary for manta rays, prioritising their management under Ministry Regulation No.4/2014, providing full protection based on Constitution No.5/1990. Additionally, Raja Ampat has designated the area as shark and manta ray sanctuaries, following the declaration of Raja Ampat Government No.9/2012 focusing on ray and shark protection. The government supports the sustainable manta ray tourism policy to boost the local economy and species conservation. According to UNWTO, tourism can play a significant role in wildlife conservation when managed responsibly and sustainably.

2.3 Manta Ray Watching Tourism

Manta rays, as charismatic marine species, symbolise the inherent value of the oceans. Their charismatic nature contributes to the economic growth of marine wildlife tourism, as tourists are drawn to experience the animal's presence (Gallagher et al., 2011; Hani et al., 2020; Krüger, 2005). Numerous scientific studies have demonstrated the industry's benefits, encompassing conservation efforts and revenue generation (Cater & Cater, 2007; Higham & Lück, 2007). In a study conducted by O'Malley et al. (2013), manta ray tourism refers to recreational activities undertaken to observe these creatures in their natural habitat, including diving, snorkelling and boat-based observation. O'Malley also researched the economic value of manta tourism in various locations, revealing a total of 139,594 manta dives with an estimated expenditure of $10,655,022. Several global manta rays sighting hot spots exist, including the Caribbean, Indian Ocean, Western Pacific and Central and Eastern Pacific, with Indonesia featuring in Nusa Penida, Komodo and Raja Ampat. In Raja Ampat, a single live manta ray is estimated to be worth up to $1 million, whereas its value drops to $500 when dead, as Conservation International (2014) reported. Prabuning et al. (2015) conducted an economic valuation study of manta rays and sharks in Komodo National Park, estimating a total value of $1.2 billion annually. The figures highlight substantial economic advantages, and it is crucial to maintain these economic benefits with sustainable practices, prioritising environmental conservation, community involvement and the welfare of local residents. A well-balanced approach guarantees the attainment of economic gains without jeopardising the destination's long-term sustainability.

2.4 Sustainable Tourism

According to the World Tourism Organization (2024), sustainable tourism is a form of tourism that thoroughly considers its present and future economic, social and environmental effects. It aims to meet the requirements of visitors, the industry, the environment and the host communities acknowledging the interconnectedness of these factors. It is an approach to tourism that seeks to maximise positive impacts while minimising negative effects on the environment, society and economy. Wood's (2002) previous study confirmed that the theory of sustainable tourism mandates holistic development of the area, which integrates the sustainability aspect of the economy, environment and socioculture. The goal is to ensure that tourism activities are conducted in a way that preserves a destination's natural and cultural resources, supports local communities, and provides a positive experience for both visitors and hosts. Many tourism destinations have recognised the importance of implementing sustainable tourism development, including marine-related tourism (World Tourism Organization, 2024). The planning and decision-making process should actively require local resident involvement to balance conservation and welfare (Blackman et al., 2004). Furthermore, Fennell and Dowling (2003) emphasise that the existence of committed institutions and individuals through effective policies and strategies is essential in order to maintain the balance between local communities, pristine areas and cultural values. Sustainable tourism is now widely promoted as a holistic concept that aims to integrate social, economic and cultural policies to ensure high-quality growth in the context to which it is applied. During implementation, such programs often encounter place-specific institutional and cultural barriers that are usually not recognised in the planning stages (Mbaiwa, 2005). Teh and Cabanban (2006) note that many developing countries face economic, social and physical challenges that pose great hindrances to sustainable development, including wildlife tourism (Hani et al., 2019a, 2019b; Lück, 2007).

3. Methodology

Cooper (1993) described the dynamic and comprehensive nature of the carrying capacity concept, wherein the number of tourists generates environmental, social and economic impacts that resource users can endure for sustainability. This notion is underscored by Mathieson and Wall (1982), who stress the importance of a tourist area's ability to fulfil tourist needs without altering environmental conditions, thereby maintaining tourism quality. This study utilised a quantitative methodology for processing tourism carrying capacity, encompassing data collection and analysis methods. The study comprises three stages: preparation, data collection and data analysis. The preparation steps involve identifying study locations, gathering necessary field materials and reviewing reports for preliminary information on manta ray tourism sites. The study locations include Dampier Straits, comprising six manta ray tourism sites (Yefnabi, Way Island, Dayan, Manta Sandy, Manta Ridge and Blue Magic). The data collection was carried out in two phases in 2018 and 2021, and this stage utilised a range of data collection methods to fulfil the study's requirements. Primary data collection involved obtaining information directly from the field through methods such as observation, measurement, calculation and interviews to gain an overview. Additionally, secondary data collection involved obtaining reports and findings from prior studies for supplementary information. Observational data collection was also conducted to gather data from the observation area, recording locations and capturing field images using a camera. Interview-based data collection was conducted to interact with targeted respondents and gather specific data or information as required. There are several formulas used to analyse data. The carrying capacity of a region denotes its capability to sustain visitors engaging in activities up to maximum thresholds without causing disturbance or deterioration in environmental quality (Fandeli, 2009). This study utilises various methods to analyse the carrying capacity of the manta ray site, including Limits of Acceptable Change (LAC) by Prosser (1986), Physical Carrying Capacity (PCC) by Fandeli (2009) and Real Carrying Capacity (RCC) Zacarias et al. (2011).

4. Result

4.1 Socio-economic Carrying Capacity

Results obtained from the data collection of community tourism service providers from villages of Arborek, Kurkapa, Kapisawar and Sawingray. Interviews were undertaken with homestay service providers, artisans, tour operators, diving operators and homestay/resort workers. The total includes 26 homestays, 2 dive operators, 9 tour operators, 30 craftsmen and 51 villagers involved in other aspects of the tourism industry. Table 3.1 shows the number of community tourism services in manta ray tourism-related businesses. Arborek has the highest number of homestays and crafts. Regarding service workers, roles such as cooks, cleaners, drivers, guides and cadres are included. The highest volume of service workers is observed in Kurkapa, with the majority employed by resorts. Sawingray, on the other hand, has the most tour operators, with each homestay offering tours and transportation services. Moreover, Table 3.2 indicates a monthly financial profile received by each community that is involved in manta ray tourism-related businesses. Earnings from service providers reflect a notable sum, with the average income demonstrating a relatively high value, enabling

Table 3.1. Number of Tourism Service Providers by Community.

Village	Homestay	Dive Operator	Tour Operator	Handicraft	Others
Arborek	12	1	2	15	18
Kurkapa	2	0	0	10	30
Kapisawar	5	0	0	5	1
Sawingray	7	1	7	0	2

Table 3.2. Average Financial Profile Per Month.

Type of Service	Income (USD)	Business Operational Cost	Household Productive Cost
Homestay	300–1,900	30–130	30–300
Dive operator	2,000–4,500	950–1,500	130–600
Tour operator	600–900	300–400	30–60
Handicrafts	30–60	3–10	10–25
Others	60–200	3–15	25–60

service providers to cover their operational and productive expenses. In this context, the survey documented various assets encompassing physical, social, humanitarian and individual assets (refer to Table 3.3). Table 3.3 outlines the ownership of assets, with varying amounts yet similar types of assets. The community expresses excellent satisfaction with the progress of manta ray tourism over the past year, as it has yielded numerous positive benefits for both village and individual development.

The findings indicate manta ray tourism has empowered local communities by providing opportunities for entrepreneurship. Locals established small businesses, such as tour operators and homestays, and sold local products, allowing them to actively participate in and benefit from the tourism industry. The results highlight the financial resources and social circumstances of the local community engaged in manta ray tourism services. This shows that the positive effects of manta ray tourism on local communities are evident in the economic benefits it brings. The creation of employment opportunities, particularly for women engaged in homestay services, artisanal crafts, tour operations, diving operations and resort-related work, contributes significantly to the community's economic growth. The income generated helps address unemployment and enhances residents' overall quality of life. Additionally, the revenue from tourism plays a role in preserving manta rays, as locals produce manta ray-shaped crafts, creating economic incentives to safeguard these assets for sustained interest from visitors. Additionally, the success of manta ray

Table 3.3. Inventory Assets Owned by Service Providers.

Physical Asset	**Sea Transport**	**Wooden Boat**
		Boat engine
	Electronic	TV
		Refrigerator
		Fan
		Mobile phone
		Sound system
		Television satellite
	House	Bricks wall
		Ceramics floor
		Tin roofing
Financial asset	Saving	
	Gold/pieces of jewelry	
Social asset	Government and NGO's fund	
	Association of tourism providers/services	
	Church congregational gatherings	
	Youth association	
	Women working group	
	Indonesian Tourist Association	
Humanity asset	Insurance	
	Higher level of education	
	Workshop and trainings by tourism stakeholders	
Individual asset	Confidence	
	Pride	
	Self-motivation	

tourism indicated community involvement and collaboration that foster social cohesion as residents work together to ensure the positive development of their community as a tourism destination. Moreover, the positive benefits establish a sense of pride in local communities, leading to efforts to conservation of the manta rays.

4.2 Limits of Acceptable Change (LAC)

Following interviews with 14 respondents, we collected findings on preferred conditions and perceived effects of manta ray tourism. The main issues that arise include the necessity for reliable monitoring posts, reducing the number of visitors to less than 20 people per hour, and installing moorings at exit and entry points (see Table 3.4 and Fig. 3.2). Moreover, the respondents propose three alternate locations for visitor distribution (Fig. 3.3). The result of manta ray sightings in Table 3.5 provides further details that illustrate the opportunity for visitors to do interaction activities. Secondary data from satellite tags reveal the consistent presence of manta rays throughout the year, with varied daily appearances aligning with high and low tides for daily feeding and cleaning activities of the manta rays. The oceanic manta rays mainly appear at Blue Magic, while other locations attract coral manta rays. This result confirms that tourist activities must be tailored to manta ray activities, whether specific locations are designated for snorkelling, diving or both. The findings from this LAC study are highly valuable for effectively managing manta ray tourism. This involves identifying acceptable environmental impacts resulting from 'desirable' social activities.

The LAC in this study refers to the acceptable level of visitor activity in manta ray tourism sites without negatively impacting the manta rays and the well-being of the locals. The result will serve as a baseline to show how to manage the social carrying capacity, allowing destinations to minimise negative social impacts on

Table 3.4. Conditions and Acceptance of Manta Ray Tourism Impacts.

Impact	Issue	Indicator	
1. Environment	1. Pollution	20%	1. Waste
		30%	2. Flying sand
2. Social	2. Conflicts	60%	3. Visitors get off at any place
		30%	4. Disturb manta rays
		55%	5. Free diving
	3. Density of visitors	90%	6. Too dense (20 people in 1 hour)
3. Psychology	4. Service	70%	7. No fresh and drinking water
		85%	8. No microphone
		80%	9. *Hospitality* (kurang ramah)
		75%	10. Limited English
	5. The monitoring post	80%	11. Not clean
		95%	12. Need renovation
		85%	13. No moorings (disappeared)
	6. Information	80%	14. Rules need to be reinforced
		60%	15. Licence check
		70%	16. Limited to swimmers

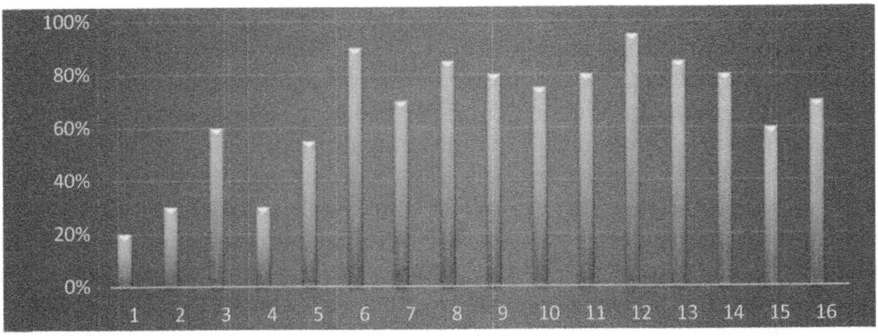

Fig. 3.2. Percentage Graph of the Manta Ray Tourism Site Issue Level.

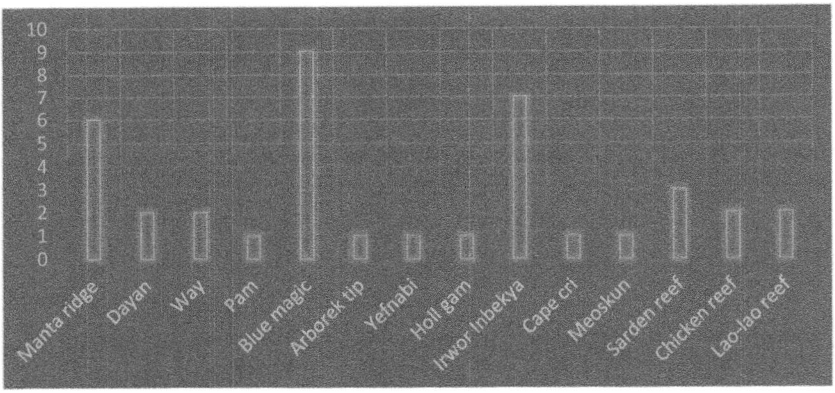

Fig. 3.3. Alternative Manta Ray Tourism Sites.

the species. Controlling tourist behaviour by considering social carrying capacity offers several benefits to preserving manta rays. The study findings indicate some alternative sites for distributing the tourists into diverse locations to prevent overcrowding and control their behaviour. This approach will improve the quality of the visitor experience and contribute to a more enjoyable and satisfying experience for tourists. While tourism can bring economic benefits, managing social carrying capacity ensures that these benefits are balanced and sustainable. It prevents situations where the economic gains from manta ray tourism outweigh the negative impacts on the species and local community. Preventing the negative consequences of excessive tourism contributes to destination resilience, ensuring the manta ray sites remain attractive and sustainable.

Table 3.5. Profile of Manta Rays in the Study Areas.

No	Location	Area of Utilisation	Species	Sightings	Time of Sightings (Daily)	Number of Manta Rays (Daily)	Interaction Activities
1	Yefnabi	Feeding and cleaning	Reef manta	All year round	Between tide to tide	1–15	Snorkelling
2	Way	Feeding and cleaning	Reef manta	Nov–April	7 a.m.–10 a.m. 2 p.m.–6 p.m.	1–13	Diving, snorkelling
3	Dayan	Cleaning	Reef manta	All year round	7 a.m.–10 a.m. 2 p.m.–6 p.m.	1–15	Diving, snorkelling
4	Manta sandy	Cleaning	Reef manta	All year round	7 a.m.–10 a.m. 2 p.m.–6 p.m.	1–6	Diving, snorkelling
5	Manta ridge	Feeding and cleaning	Reef manta	All year round	7 a.m.–10 a.m. 2 p.m.–6 p.m.	1–15	Diving
6	Blue magic	Cleaning	Oceanic manta	Nov–April	7 a.m.–10 a.m. 2 p.m.–6 p.m.	1–3	Diving

4.3 Physical Carrying Capacity

The available space for the movement of mantas varies significantly at each study location, with the Blue Magic being the smallest. Calculations for tourist utilisation consider the species' area (cleaning/feeding), and the sensitivity of the utilisation area for the manta rays. Tourist rotations per day are determined based on the daily appearance hours of manta rays, with a set limit of five rotations. The required space for each snorkelling tourist or diver is 500 m^2. The applied limits per hour for divers and swimmers range from 10 to 15, with detailed calculations in Table 3.6. The results of the physical carrying capacity represent an emerging field for establishing the optimal number of manta ray tourists in each site. The results indicate the ability of the sites to tolerate and adapt to tourist numbers and avoid overcrowding.

As a baseline study, the findings are important to ensure that manta ray tourism activities do not degrade the sites from overcrowding. Adhering to physical carrying capacity helps protect the manta rays by preventing overuse, overcrowding, habitat destruction and other negative impacts that align with the sustainable tourism concept. Establishing and enforcing limits on visitor numbers supports the long-term sustainability of the destination and manta ray conservation while ensuring balance and satisfaction for those engaged in manta ray tourism activities, both visitors and the local community. Moreover, adjusting carrying capacity based on research findings allows for adaptive management practices that align with conservation goals. Physical carrying capacity is a key component of sustainable and responsible tourism practices supporting the conservation objectives of the manta rays.

5. Conclusion

The connection between manta ray conservation and tourism is intricate, and achieving a delicate equilibrium is essential for sustainable development. While tourism can offer constructive support to conservation through financial aid, awareness-building and emphasising the significance of natural and cultural assets, there is also the potential for tourism to present challenges and risks to conservation if not appropriately managed. Successful conservation-tourism collaborations demand meticulous planning, stakeholder engagement and the adoption of responsible practices to guarantee that tourism's advantages actively contribute to preserving the manta rays.

6. Implication

This study possesses unique characteristics, necessitating site management to adapt accordingly. The application of findings from carrying capacity studies is pivotal and demands extensive coordination for the development and planning of the associated management strategy, ensuring its optimal execution. Sustainable manta ray tourism entails adopting measures and programs to mitigate adverse impacts on manta ray populations and their habitats while concurrently fostering conservation, education and responsible tourism. Manta rays, charismatic marine

Table 3.6. Results of the Study of the Physical Carrying Capacity of the Study Areas.

No	Location	Area of Movement (Manta Rays)	Area of Visitor Intercations	Rotational Coefficiency	Movement Space Needs (Per Individual)	PCC	RCC	Number of Visitors Per Hour
1	Yefnabi	2.745.504	11.098	5	500 m^2	111	74	15
2	Way	3.615.204	11.286	5	500 m^2	113	76	15
3	Dayan	252.428	8.177	5	500 m^2	82	55	11
4	Manta sandy	17.546	7.503	5	500 m^2	75	50	10
5	Manta ridge	112.337	10.403	5	500 m^2	104	70	14
6	Blue magic	5.475	5.475	5	500 m^2	55	37	7

creatures that attract tourists for activities like snorkelling and diving, require a set of recommendations for sustainable practices. This includes collaborative efforts among stakeholders, encompassing community members, local authorities, tour operators and dive centres, to formulate and implement strategies that mutually benefit the local economy and manta ray conservation.

7. Recommendation

To ensure sustainable manta ray tourism, stakeholders are advised to engage in continuous research and monitoring, enforce guidelines and support conservation initiatives. This collaboration aims to preserve manta rays while facilitating responsible tourism experiences. Active participation from local communities is crucial, involving them in conservation efforts to develop a sense of ownership and responsibility for protecting manta rays and their habitats. Stakeholders can further contribute by supporting local manta ray conservation initiatives through financial aid, volunteering or collaboration with organisations that safeguard the species. Educational programs targeting both tourists and local communities play a vital role in raising awareness about the significance of manta rays, their conservation status and the role of tourism in protecting them.

In implementing sustainable tourism practices specific to manta ray tourism, stakeholders should focus on using environmentally friendly equipment, minimising waste and adhering to eco-friendly practices to reduce the overall impact on the marine ecosystem. Developing and enforcing guidelines and regulations, such as restricting tourist numbers, defining safe viewing distances and implementing codes of conduct, are essential to prevent animal disturbances. Additionally, encouraging responsible wildlife interactions and emphasising passive observation over activities like touching or feeding helps maintain the natural behaviour of manta rays and avoids causing stress. Seasonal restrictions aligned with critical life stages, such as avoiding certain areas during mating or pupping seasons, further minimise disturbances and contribute to sustainable manta ray tourism.

8. A Set of Sustainable Practices for Manta Ray Tourism

8.1 Technical Guidelines

- Implement the study findings for effective tourist management on-site.
- Facilitate discussions between the area management authority and the local community to review study results and develop site management plans.
- Create infographics for each site to provide technical guidance.
- Conduct community training, including learning from manta cadres at post, instructing on ID photo capture and collecting visitor registration data.
- Optimise existing infrastructure, incorporating educational interpretation media. Provide another floating post as a visitor monitoring station, the land post as a supporting monitoring post and a tourist information centre.
- Ensure compliance with Cleanliness, Health, Safety, and Environmental Sustainability (CHSE) certification standards for posts and information centres.

8.2 Ethical Guidelines

The ethical guidelines outlined here promote responsible conduct among area users, particularly during manta ray observation activities, based on principles adopted from the Manta Trust.

- Maintain 10 metres during observation.
- Adhere to speed limits of 8 knots within 100 metres and 5 knots within 30 metres.
- Avoid passing through manta ray cleaning stations.
- Prohibit boats from carrying diving/snorkelling tourists above specified stations.
- Ensure tourists do not swim or snorkel through designated station locations.
- Prohibit boats and tourists from disrupting the movement or cornering of rays.
- Limit the duration around the biota to a maximum of 60 minutes.
- Strictly avoid physical contact, including touching, riding or feeding rays.
- Utilise environmentally safe sun lotion to prevent irritation to the biota.

8.3 Waste Management

All area users implement and follow daily regulations to support effective waste management.

- Instal warning and prohibition boards at strategic locations, emphasising the importance of protecting the land and sea environment and indicating the location of rubbish bins.
- Promote the 'Hero Waste' programme, encouraging tourists to actively participate in waste reduction during their trips.
- Task the patrol team with garbage collection.
- Implement an extensive recycling programme to minimise the environmental impact of waste.
- Consider providing drinking water filling stations to reduce the use of plastic or glass bottled mineral water.
- Encourage tourism businesses to independently manage their waste without leaving any in the area.

8.4 Anchoring

Prepared anchor locations are designated for recreational vessels engaging in activities or overnight stays. Anchoring is only allowed in predetermined areas, communicated through location maps and GPS coordinates on the website and information system, with marked anchor drop zones using navigation buoys.

8.5 Mooring

Placing moorings for maritime tourism requires initial studies and collaboration with transportation and other partners. Key steps include.

- Identifying user groups.
- Assessing mooring requirements, considering factors such as location, substrate type and depth.
- Identifying material suppliers.
- Selecting suitable mooring systems (e.g. Halas, helix, hughes, bewr, manta ray).
- Providing training for installation and maintenance.
- Disseminating information to user groups.
- Once installed, mooring buoys should be marked with code stickers in distinct colours. These colours serve as indicators for ships to understand the strength limits of moorings under various conditions, such as during strong winds. See Table 3.7 for an example of the colour-coded indications for mooring buoys.

8.6 Regular Tasks

- Replace damaged buoys and signs as part of routine maintenance.
- Conduct scheduled mooring maintenance three times a year, documenting activities in a systematic logbook.
- Maintain a stock of mooring maintenance supplies, including ropes, reels, buoys, sekels, horse hooves, chains and airbags of various sizes.
- Allocate a budget for fuel, tanks, diving equipment and logistics for monitoring and maintenance activities.
- Undertake routine monitoring and maintenance of moorings.

8.7 Rules for Mooring Use

- Adhere to restrictions, including time limits, structural capacity and other conditions specified on the mooring buoy or sticker.

Table 3.7. The Colour-Coded Indications for Mooring Buoys.

Colour	Class	Wind Strength Limit	Maximum Boat Length (Monohulls)	Maximum Boat Length (Multihulls)
Chocolate	Tender	24 knots	6 metre	6 metre
Yellow	Class A	24 knots	10 metre	9 metre
Green	Class B	34 knots	20 metre	18 metre
Blue	Class C	34 knots	25 metre	22 metre
Red	Class D	34 knots	35 metre	30 metre

- Allow only one vessel to use a mooring at a time, except for tender vessels attached to the main vessel.

8.8 Physical Development Guidelines

- Regulate the construction of jetties, permitting only MPAs to build in dense tourist areas and high ecosystems, using environmentally friendly materials.
- Prohibit other physical constructions supporting tourism, including accommodation/resorts in the area.
- Subject additional infrastructure development approved in the master plan to an environmental assessment for justification.

8.9 Tourist Boats Must Adhere to the Following Standards

- Possess business and operational legality.
- Have life jackets in good condition, suitable in size and quantity for the crew and passengers.
- Equip anchors, chains/lines, bailer/bucket/spray bucket for fire, bilge pumps, compass, smoke flare, fire extinguisher, maps, paddle, first aid, rescue sled, two-way communication device, sound signals, tool kits, floating tow rope, bow crane, interpretation materials, waste bins and separate waste disposal for organic and inorganic waste.
- Conduct regular training, checking and evaluation for waste handling by the ship's crew.

Prohibit smoking on board.

8.10 Information Boards

- Location map: Display tourist locations, boundaries, facilities and routes, providing geographical and administrative information.
- Interpretation board: Offer details about charismatic biota and natural resources in protected areas, along with best practice guidelines.
- Compliance board: Emphasise the principle of 'Take only photographs and leave only footprints'.

8.11 Tourist Information System

- Socialise and provide access to the information above through a website, accompanied by explanations of various user needs.

In addition, illustrations 1–6 outline each site's profile, which has been fine-tuned according to the data collection and processing findings.

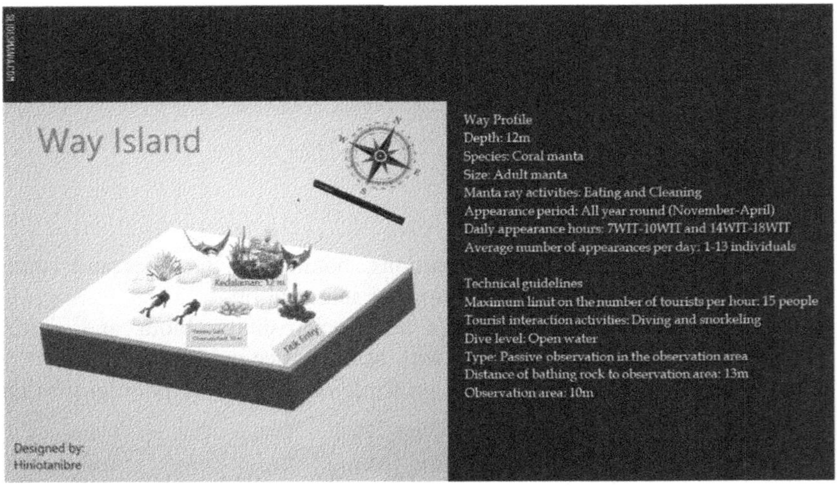

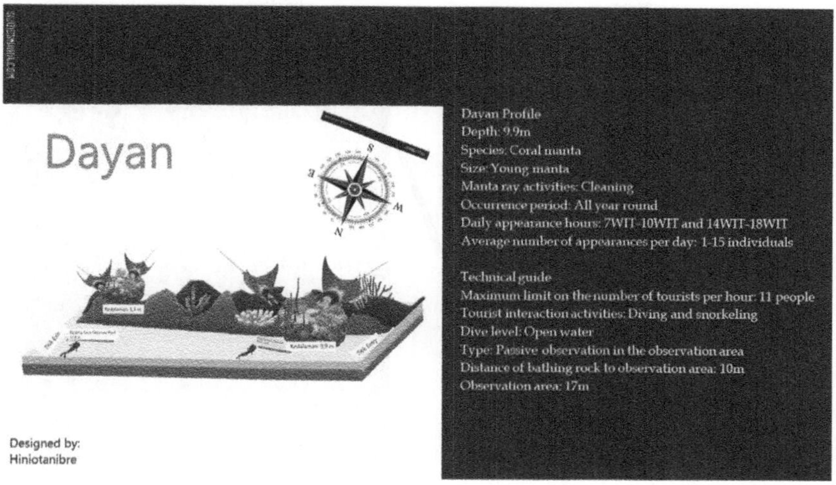
Dayan
Kedalaman: 9.9 m
Designed by:
Hiniotanibre
Dayan Profile
Depth: 9.9m
Species: Coral manta
Size: Young manta
Manta ray activities: Cleaning
Occurrence period: All year round
Daily appearance hours: 7WIT-10WIT and 14WIT-18WIT
Average number of appearances per day: 1-15 individuals
Technical guide
Maximum limit on the number of tourists per hour: 11 people
Tourist interaction activities: Diving and snorkeling
Dive level: Open water
Type: Passive observation in the observation area
Distance of bathing rock to observation area: 10m
Observation area: 17m

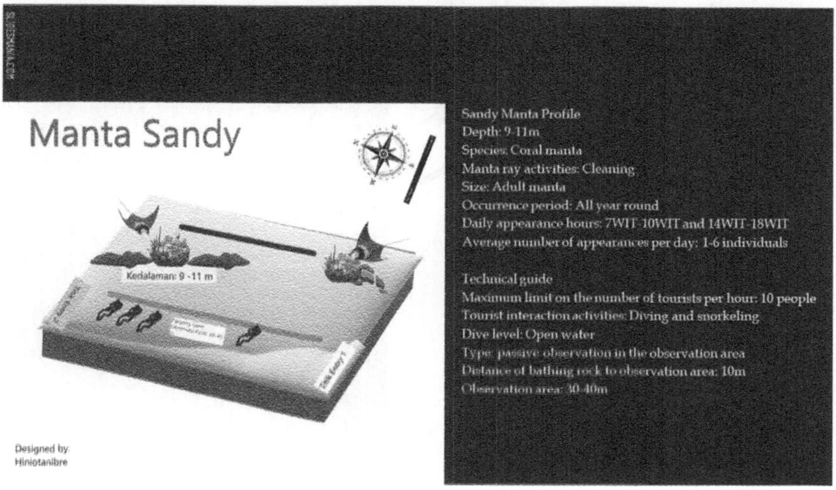
Manta Sandy
Kedalaman: 9 -11 m
Designed by:
Hiniotanibre
Sandy Manta Profile
Depth: 9-11m
Species: Coral manta
Manta ray activities: Cleaning
Size: Adult manta
Occurrence period: All year round
Daily appearance hours: 7WIT-10WIT and 14WIT-18WIT
Average number of appearances per day: 1-6 individuals
Technical guide
Maximum limit on the number of tourists per hour: 10 people
Tourist interaction activities: Diving and snorkeling
Dive level: Open water
Type: passive observation in the observation area
Distance of bathing rock to observation area: 10m
Observation area: 30-40m

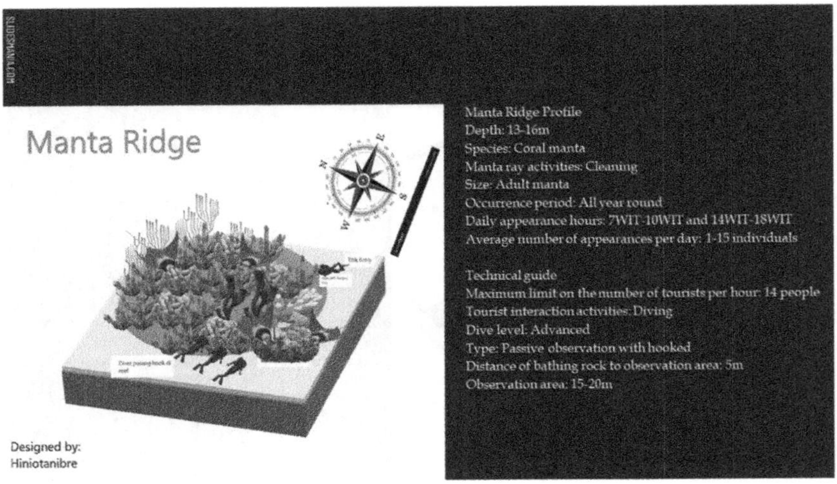

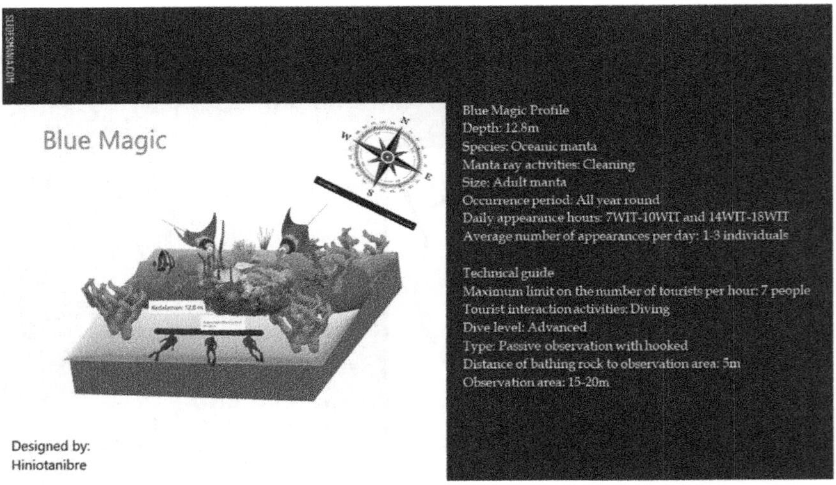

Acknowledgements

This research owes its feasibility to Konservasi Indonesia, which advocates for sustainable development and environmental conservation in Raja Ampat, Indonesia. Special recognition goes to the Elasmobranch Program of Konservasi Indonesia and the personnel with the team from BLUD UPTD Raja Ampat, who actively participated in data collection. We express our heartfelt gratitude to all stakeholders for their cooperation in this study.

References

Blackman, A., Foster, F., Hyvonen, T., Kuilboer, A., & Moscardo, G. (2004). Factors contributing to successful tourism development in peripheral regions. *Journal of Tourism Studies, 15*(01).

Cater, C., & Cater, E. (2007). *Marine ecotourism: Between the devil and the deep blue sea*. CAB International.

CITES. (2014). Sharks and manta rays. https://www.cites.org/eng/prog/shark/more.php

Cooper, C. (1993). Tourism principles & practice. Longman Group Limited.

Deakos, M. H. (2010). *Ecology and social behavior of a resident manta ray (Manta alfredi) population off Maui*. University of Hawaii.

Deakos, M. H., Baker, J. D., & Bejder, L. (2011). Characteristics of a manta ray Manta alfredi population off Maui, Hawaii, and implications for management. *Marine Ecology Progress Series, 429*, 245–260.

Dewar, H., Mous, P., Domeier, M., Muljadi, A., Pet, J., & Whitty, J. (2008). *Movements and site fidelity of the giant manta ray, Manta birostris in the Komodo Marine Park*. Springer-Verlang Marine Biology.

Fandeli, C. dan M. (2009). *Prinsip-prinsip dasar mengkonservasi landskap*. Gadjah Mada University Press.

Fennell, D. A., & Dowling, R. K. (2003). *Ecotourism policy and planning*. CABI Publishing.

Gallagher, A. J., Lazarre, D. M., & Hammerschlag, N. (2011). Global shark currency: The distribution, frequency, and economic value of shark ecotourism *Current Issues in Tourism, 14*, 797–812.

Hani, M. S., Jompa, J., Nessa, M. N., & White, A. T. (2019a). Manta ray watching tourism in Eastern Indonesia: Is it sustainable? *IOP Conference Series: Earth and Environmental Science*.

Hani, M. S., Jompa, J., Nessa, M. N., & White, A. T. (2019b). Behavioural analysis of manta ray tourists in Eastern Indonesia. *IOP Conference Series: Earth and Environmental Science*.

Hani, M. S., Lück, M., Jompa, J., Nessa, N. M., White, A., & Hanifah, M. I. H. P. (2020). The role of small-scale manta ray tourism businesses in Eastern Indonesia. *International Journal of Innovation, Creativity, and Change, 12*(3), 89–102.

Higham, J., & Lück, M. (2007). *Marine wildlife and tourism management: Insights from the natural and social sciences*. CABI.

IUCN. (2011). Red list of threatened species. http://www.iucnredlist.org

Krüger, O. (2005). The role of ecotourism in conservation: Panacea or Pandora's box? *Biodiversity & Conservation, 14*, 579–600.

Lück, M. (2007). *Marine wildlife and tourism management*.

Marshall, A. D., Pierce, S. J., & Bennett, M. B. (2009). Morphological measurements of manta rays (Manta birostris) with a description of a foetus from the east coast of Southern Africa. *Zootaxa, 30*, 24–30.

Mathieson, A., & Wall, G. (1982). *Tourism: Economic, physical, and social impacts*. Longman.

Mbaiwa, J. E. (2005). The problems and prospects of sustainable tourism development in Okavango Delta. Botswana. *Journal of Sustainable Tourism, 13*(03), 203–227.

Notarbartolo-di-Sciara, G., Serena, F., & Mancusi, C. (2006). Mobula mobular. IUCN red list of threatened species. http://www.iucnredlist.org/apps/redlist/details/39418/0

O'Malley, M. P., Lee-Brooks, K., & Medd, H. B. (2013). The global economic impact of manta ray watching tourism. *PLoS One*, *8*, e65051.

Prabuning, D., Setiasih, N., Priyantoro, A., & Harvey, A. (2015). *Management alternative of shark and manta tourism: A case study of economic valuation Simposium Hiu dan Pari di Indonesia.*

Prosser, P. (1986). The limits of acceptable change: An introduction to a framework for natural area planning. *Australian Parks and Recreation*, *22*(2), 5–10.

Teh, L., & Cabanban, A. S. (2006). Planning for sustainable tourism in southern Pulau Banggi: An assessment of biophysical conditions and their implications for future tourism development. *Journal of Environmental Management*, *85*(04), 999–1008.

Wood, M. E. (2002). *Ecotourism: Principles, practices and policies for sustainability.* United Nations Environmental Program.

World Tourism Organization. (2024). *Sustainable development of tourism conceptual definition.* http://www.world-tourism.org/sustainable/concepts.htm. Accessed on 22 January 2024.

Zacarias, D. A., Williams, A. T., & Newton, A. (2011). Recreation carrying capacity estimations to support beach management at Praia De Faro, Portugal. *Applied Geography*, *31*, 1075–1081.

Chapter 4

The Transformative Aspects of Green Tourism: A Himalayan Perspective

Arun Bhatia[1], *Aditi Sharma*[2] *and Ruchi Sharma*[3]

Central University of Himachal Pradesh, India

Abstract

The Fourth Industrial Revolution is creating a substantial buzz, yet there is another potentially transformative force which is making its appearance in the form of "green" revolution. Scientists, Social Scientists and Environmentalists worldwide are leveraging platforms to amplify their messages on protesting against the evils of anthropocentrism. Thus, in the realm of business, there's a noticeable shift towards focusing on green production, green services and green consumption. This trend has given rise to a new kind of entrepreneurship known as green entrepreneurship, which emphasises the creation of sustainable businesses with an environmentally conscious outlook. Eco-preneurs or eco-capitalists are being hailed for their noteworthy efforts in adopting green practices and promoting sustainability. An illustrative instance of this trend is Green Tourism, where tourists are nudged towards low ecological footprint leisure. In this chapter, the authors delve into the case studies of green entrepreneurs in the Himalayan region, examining how their initiatives have brought about a profound revolution in the lives of the local communities.

Keywords: Green tourism; eco-preneurs; sustainable tourism; ecotourism; anthropocene challenges

[1]https://orcid.org/0000-0003-3464-6128
[2]https://orcid.org/0000-0001-5774-528X
[3]https://orcid.org/0000-0003-0995-1876

Sustainable Tourism, Part B, 67–84

doi:10.1108/978-1-83608-326-920241004

1. Introduction

Ensuring sustainable development and tourism is crucial in the face of the Anthropocene challenges and irreversible environmental changes brought about by human activity. Both global and national leaders are calling for the achievement of the Sustainable Development Goals (SDGs) and urging governments to prioritise and align with them. There is an urgent need to transition from measuring progress beyond GDP to factoring in sustainable human well-being and social equity (Higham & Miller, 2018).

> Everyone is transformed by travel to some degree. Lean (2012, p. 169) cited Sheldon (2020).

The relationship between travel and personal growth is frequently illustrated through spatial or geographic language, such as crossing thresholds, ascending or descending, to depict alterations in the non-spatial mind or consciousness. Yet, embarking on a physical voyage is also seen as an external expression and crucial impetus for an internal psychological or spiritual expedition. This twofold nature, both symbolic and factual, has been employed by diverse philosophical and religious beliefs for ages (Morgan, 2010).

Tourism industry is essentially based on the idea of experience selling to its consumers and helps the tourist to escape from the mundane hassles of life. The tourists ventures out to explore in search of newness, driven by the hedonic motives and sometimes in hope of recharging and transforming oneself. Towering Himalayas have long been visited by the saints and seers in search of deeper meaning of life and still continue to attract visitors who wish to have some cathartic life experiences. Kirillova et al. (2017a) highlighted the importance of personal development, social interaction and cultural immersion in creating transformative tourism experiences. Sheldon (2020) asserts that transformative tourism experiences can be achieved through activities that promote personal growth, meaningful connections and authentic cultural experiences.

As the most rapidly growing industry, the tourism sector has the potential to serve as a powerful force for transformation, igniting the process of inner change and resulting in heightened awareness (Bhalla & Chowdhary, 2022). Green tourism is emerging as a buzz word as more and more environmentally conscious tourist are preferring the business ventures that have adopted eco-friendly measures. These businesses not only attract the customers but are also able to retain them by selling the 'green concept'. In the global service industry, tourism industry contributed 10.2% of the world's GDP and generated 1 out 10 jobs in the world's job market (Siyambalapitiya et al., 2018). The tourism industry has undergone significant changes due to the Covid-19 pandemic. To ensure the preservation of the environment and local cultures, it is imperative to revamp the tourism ecology. As Gill et al. (2019) highlights, addressing the challenges that impact people and the environment are key to transforming the industry. A promising avenue to achieve this is through green tourism transformation, which entails embracing eco-friendly business models and multi-impact frameworks.

Borel-Saladin &Turok (2013) emphasised 'green economy has the potential to effect substantive and transformative change towards the goal of sustainable development'.

Recent research suggests that a fresh perspective is needed on sustainable and resilient tourism. By fostering a greater consciousness and awareness, tourists can actively contribute to the restoration and regeneration of local ecosystems, enhancing their health and resilience to climate change (Wahl, 2019). Concurrently, tourism can promote personal and social transformation, ultimately improving the well-being of individuals (Kottler, 1998; Ross, 2010). Transformative travel can forge an emotional connection between people and nature, resulting in significant personal growth (Ateljevic, 2020; Sheldon, 2020). To ensure the long-term sustainability of tourism and visitor economies, it is crucial to rethink operational systems within the industry (Cave & Dredge, 2020). Studies have emphasised the importance of regenerative tourism, which involves implementing targeted initiatives and approaches to promote the overall sustainability and well-being of both communities and individuals (Ateljevic, 2020; Bhalla & Chowdhary, 2022; Duxbury et al., 2020).

The multifaceted nature of the tourist experience, coupled with various points of contact with tourism services, renders designing for any form of tourism a formidable endeavour (Tussyadiah, 2014). Crafting a tourism experience that aims to facilitate inner transformation is a challenging task that demands meticulous attention to detail and subtlety (Sheldon, 2020).

India has become a sought-after destination for travelers seeking a profound and transformative experience, allowing them to reconnect with their authentic selves and simply exist in the present moment. The Himalayan region, in particular, is well-suited for such an experience, providing a refreshing atmosphere, natural surroundings, suitable weather, healing properties and the opportunity for forest bathing.

Travel with a purpose of transformation makes us more empathetic, especially when we interact with nature (Sheldon, 2020). The sense of oneness, peace and connection with the nature causes us to grow as individuals. In the modest way possible little research has been presented in literature on the transforming experiences associated with green tourism, and there may be very few coming from rising economies such as India. This chapter, through several case studies, explains how eco-preneurs in the Himalayan region may go a long way in promoting sustainability by adopting practices that can bring about life-changing shifts further leading to positive self-improvement in individuals.

2. Theoretical Framework

2.1.1 Explanation of Key Concepts: Green Tourism, Sustainability and Transformative Aspects

The United Nations Environment Programme (UNEP) and World Tourism Organization (UNWTO) jointly released a report titled "Tourism in the Green Economy" that highlighted the significance of investing in eco-friendly and sustainable tourism as a means to generate employment opportunities and reduce poverty, while also contributing to environmental preservation (United Nations Environment Programme and World Tourism Organization, 2012). Meeting up the future challenges from tourism, Furqan et al. (2010) asserts green as a concept in tourism that is easy to define but difficult to practice. Rooting back to the concept of rural tourism Yamazaki (1996) highlighted green tourism as a mechanism has paved the roadmap for future tourism where demands and supply gap is met overcoming the challenges of destination marketing and destination promotion. Walker and Hanson (1998) provide evidences on green tourism for building blocks for strong brand equity destinations. In a nutshell, green economy revolves around two major premises that is the production of the products or services that are green in nature and secondly, design of products and services showing responsible behaviour towards the environment concern and different interest groups. Thus, green tourism is defined as a practice of tourism which rests on four major pillars of sustainable development goals in which efforts are carried out to maintain and sustain the tourism. Buhalis (2022) critically acclaim the point that societies should be out rightly paid off for the resources which are used for tourism as an economic activity in the region and work to produce co-created value, sustaining the position of all the stakeholders fairly.

With the voices being raised from social scientist and environmentalist worldwide to raise up the concern and provide ample platforms and forums to make stakeholders address issues like climate change, draughts and excessive heating in some areas, loosing of the soil, discharge of water level are alarming signs both for policy planners and implementing agencies. Based on reporting and accepting the paradigm shift in choice behaviour of customers, it has further increased the interest and friendly behaviour towards environment by the entrepreneurs. Allen and Malin (2008) defines new set of modern entrepreneurs which can be critically identified with key characteristics like less interest in the economic motive of business, more knowledge orientation towards impact of business on the environment, showing concern towards the social justice, self-motivated with a key mission, giving importance to the concept of locality and keen eye for forward thinking with sustainable behaviour and practices. In continuation to the motto of sustainability and an effort to attain the maximum from the key underpinnings of the choice behaviour of the tourists and service providers travel and tourism sector has shown transformation change in adapting and accepting the shifts in tourism sector, government agencies, tourists and lastly environment as a whole. Hall et al. (2020) with the comprehensive evaluation of the literature on the outcome of Covid-19 and its impact on the travel and tourism industry in the modest way possible tried to highlight that the goals from sustainable tourism is only possible where sustainable tourism is imbibed as global approach within an environment limits. Buhalis (2022) adding on the same highlights transformation in tourism should lead to selection of smart strategies, critical evaluation of mechanism to support sustainable development and inclusive societies.

Modern approaches to sustain the ethnicity of culturally rich destinations and further providing benefit can fruitfully be achieved while creating and providing stages for local economies to showcase richness and uniqueness of culture along with drawing inferences from entrepreneurial development in the region. Furqan et al. (2010) emphasis the tourist likes to hear, select and travel to those destinations which are termed to be neat and clean. Individual and collective efforts of entrepreneurial ecosystem at the destination are an outcome of value to tourism services. The same is judged and measured in multiple domains to make destination more competing at international levels to further attract and convert tourists for repeat visitation (Schofield et al., 2020).

Travel provides multiple opportunities to transform one, transform environment, transform business and transform economy. Coghlan and Weiler (2015) admit transformation is a personalised process that leads to a critical understanding of oneself, resulting in a new self-concept. This suggests that changes within a person may not be noticeable to others as they involve self-reflection (Coghlan & Weiler, 2015; Kirillova et al., 2017a; Snyder, 2008). In order to have a meaningful transformation as a conscious effort is necessary and transformative travel can aid the process (Boswijk et al., 2013; Kirillova et al., 2017b; Mezirow, 1991). The process of transformation entails a permanent change that fosters development and requires acquiring new knowledge to apply it effectively (Reisinger, 2013).

Although the topic of transformational experiences has received scholarly study, it is uncertain what causes a tourism encounter to become transformative. An empirical research of white water rafting experiences conducted by Arnould and Price (1993) identified three dimensions of extraordinary experiences: (1) a deeply felt connection with a physical environment, (2) a meaningful connection with fellow travelers and (3) personal growth and renewal. A comparable idea is transcendent experience, which is defined as a moment of ultimate subjective awareness, extreme enjoyment, liberation and a sense of harmony with the entire world (Kirillova et al., 2017a; Williams & Harvey, 2001).

Williams and Harvey (2001) conducted a study on transcendental experiences in forest settings and found two types: Diminutive Experiences, which involve high interest and novelty, and Deep Flow, which involve high absorption and moderate novelty. According to Levin and Steele (2005), there are two types of transcendent experiences – green and mature. Green experiences are brief, pleasurable and often ecstatic, with varying levels of intensity that include unusual subjective sensations. Mature experiences, on the other hand, are longer lasting and associated with feelings of peace and equanimity. They can result in transformative changes in one's consciousness or spiritual awareness and have the potential to positively impact a tourist's life. Thus, several studies in tourism have examined the concept of transformative experiences (Kirillova et al., 2017a) and their impact on well-being of travellers.

The transformative nature of tourism services is outlined, produced and moderated in contemporary tourism literature in multiple contexts. Buhalis (2022) strongly affirm as conventional strategies to run and operate the industry is becoming obsolete with passage of time therefore revamped action based approach is necessary to develop strong industry base meeting goals of sustainable development as set for 2030. Kirillova et al. (2017a) bringing out deep understanding to the transformation in tourism with at one end of the continuum brings transformative tourist experience and later to the same it act as a catalyst bringing subsequent long-term changes. Modern transformation is seen in high dependency of information, communication and technological use in travel value chain. The success for the destinations is an outcome to the acceptance to 3C formula in modern business ventures where first C stands for content, second C stands for community and third C stands for commerce (Bhatia, 2022). The scientific, theoretical and human centric perspective has produced a great interest in products which are green (sustainable) in nature both among products and services and brings transformative experiences. Prihanto & Kurniasari (2019) provide evidences from the results of 100 hospitality units in Indonesia and emphasise transformation in the form of digital market capabilities, digital leadership capabilities and digital technology capabilities as a prime source for digital transformation in the hospitality sector and essential for customer engagement.

2.1.2 Overview of Existing Theoretical Frameworks or Models Relevant to the Study's Context

The application of these frameworks to understand the transformative aspects of green tourism in the Himalayan region can be summarised as follows:

Dynamic Economic Model (Marsiglio, 2015): This model illustrates how tourism can drive both economic and environmental growth. In the Himalayas, sustainable tourism practices can foster economic development while preserving the environment with an aim for long-term sustainability.

Green Practices in Hotel Supply Chain (AI-Aomar & Hussain, 2017): AI-Aomar's framework assesses green practices within a hotel supply chain context. In the Himalayas, hotels play a crucial role in promoting green tourism. Implementing green practices, such as waste reduction and responsible sourcing, can enhance the sustainability efforts of Himalayan hotels.

Stimulus-Organism-Response (S-O-R) Framework (Wang et al., 2018): Wang et al.'s S-O-R framework examines the link between a green hotel's environmental image and tourists' intent to recommend such hotels. In the Himalayan region, this framework can help hotels understand how their environmental image influences tourists.

Theory of Green Purchase Behaviour (Han, 2020): Han's framework, which identifies attitude, responsibility and social norms as influencers of green purchase behaviour, is relevant to tourists in the Himalayan region. Tourists visiting this area often hold strong environmental attitudes, feel responsible for preserving the ecosystem and follow social norms promoting sustainability.

Sheldon's Connection to Environment (2020): Sheldon's research underscores the value of tourists feeling connected to the environment. In the Himalayas, tourists are drawn to the pristine natural beauty and serene landscapes, providing a unique opportunity for relaxation, self-reflection, self-awareness and consciousness.

Extended Self-Identity-Based Electric Product Adoption Model (Horner & Swarbrooke 2020): Although originally designed for the airline industry, this model's core concept of green self-identity applies to green tourism in the Himalayas. This model can help the Himalayan tourism businesses in catering to the environmentally conscious preferences of the tourists visiting these areas.

In nutshell, these frameworks provide a valuable input for examining the eco-friendly tourism in the Himalayas. Prioritising sustainable behaviours, environmental consciousness and the positive influence can not only have a positive influence on the health and preferences of visitors but also maintain the pristine beauty of Himalayas.

2.1.3 Discussion of How These Frameworks Are Applied to Understanding Green Tourism's Transformative Aspects in the Himalayan Region

Creating a tourism experience that caters to the needs of visitors can be a complex task, especially given the multitude of factors involved in the process (Tussyadiah, 2014). However, crafting a journey that is centred on personal growth and spiritual development can be even more challenging. It requires a deep understanding of the values, motivations and mindset of the traveller, as well as their readiness and openness to transformation. Moreover, while some tourists specifically seek out transformative experiences, others may come across them unexpectedly. These changes can occur suddenly or gradually over time, and can happen during any type of tourism adventure, regardless of its original purpose (Kirillova et al., 2017a; Voigt et al., 2011) thus becoming over period of time permanently. Achieving transformative experiences requires careful consideration of service encounters, touch points throughout the journey and the values, norms and philosophies guiding service interventions (Sangiorgi, 2011; Sheldon, 2020). The challenge lies in integrating these transformational moments into a lasting consciousness.

A large number of subjective assessments have been made and applied to concept of transformation in view to improve the destination and social status of local economies (Teoh et al., 2021). At one end of the continuum the green tourism expresses the intention to make a change in sustainable environment (Pan et al., 2018) to be indebted with activities engaged in tourism that are eco-friendly, low carbon foot prints and alternative tourism. Further, the same highlights bringing change in sustainable economy, sustainable culture and sustainable society. The fragile ecosystem of the Himalayan region in the Indian subcontinent provide multiple challenges, however due to the mesmerising and enchanting landscape it attracts tourist to come and visit its beautiful flora and fauna. Horner and Swarbrooke (2020) admit and provide evidence highlighting the interest of Indian domestic inhabitants of northern plains to take trips to hill stations at time of hot summer months. Within the core of influx of tourists during the peak season the transformation in the accommodation sector has provided a new direction to ventures to provide unique experiences in staying options that are unique, nature abundant and eco-friendly in showcase (Table 4.1).

Table 4.1. Synthesis of Transformative Aspects of Green Tourism.

Activities Undertaken	Socio-Economic Quality Parameters	Results Achieved
Kath Kuni Architecture	Usage of local building material, use of local art and craft, use of indigenous knowledge	Green homes, protection of local art and craft, advancing the knowledge to the next generation
Promoting EV Vehicles	Usage of EV vehicles reduces pollution, managing over-tourism at destination, encourage the usage of EV vehicles	More green initiatives by the government to promote green mobility, green support to non-government organisation's (NGO's) initiative towards environment protection
Worldwide Opportunities on Organic Farms (WWOOF)	Usage of best land usage for non-pesticides in the organic farming, attracting the foreign nationals for experiential tourism	More tourism inflow with more green tourism activity in the region, adding more employment reducing socio-economic disparity, respect to the local methods and historic methods of production
The Lodge at Wah, Palampur	Usage of landscape to promote the local environment along with tea tasting promotion to the tourists	Developing green human resource with capacity building of local community to blend the local tea production and promoting among the tourists
Rakkh Resort Palampur	Usage of farm to fork strategy applying the principles of circular economy	Developing the interest of local community towards green initiatives of the hospitality sector in the region. The same also brings experiential green tourism experience to the tourists
Geographical Indication Tags	Application of the concept for protection and	Helps in identity creation for the local community to involve in the

Table 4.1. *(Continued)*

Activities Undertaken	Socio-Economic Quality Parameters	Results Achieved
	promotion of indigenous products	production and protection of the same. The engagement of the people in the same allows for removal of socio-economic disparity gap

Source: Synthesis of literature, Developed by Authors.

The attraction of tourists in the Himalayan region to mud houses with local aesthetic sense from Kath Kuni architecture of Kullu region to mud houses in Kangra valley all provide impetus to green tourism and transformation aspects in the Himalayan region to think and protect the nature in the best possible way. Transformation is not restricted to the accommodation segment of travel and tourism rather carbon neutrality has been at the base of reduction of CO_2 emission. The government and policy intervention to reduce carbon footprints comes from mobility solution both for localities and tourists visiting the destination (Ministry of Tourism, 2022). In the same direction the Himalayan region adapted for change introduced non-pollution electric vehicles to travel to the most thought of tourist spot, i.e. Rohtang pass in Manali, Himachal Pradesh bringing the transformative experience to travel in non-polluting non-noise vehicle. The experience adds to the value to life of an individual with contribution to green environment in simplest way possible during travel journey. Along with the same the introduction of green tax with pollution certificate multiplies the contribution of government initiatives towards green mobility solution and green tourism. The acceptance to green tourism cannot be neglected with entertainment at the core of the sustainable green tourism. The Himalayan region provided ample of examples to entertain people in the local environment with creating transformation to the conventional tourism practices to more eco-friendly sustainable goal-oriented tourism. Worldwide Opportunities on Organic Farms (WWOOF) is both entertainment which is eco-friendly in nature, culture exchange, building up community with respect to nature and sustainable practices. The above-mentioned example is the perfect blend of transformative practices with an initiative to promote green tourism. One of transformative approaches to green tourism is dedicated in production and promotion of local produce which is organic in nature, highlighted in background tourism elements within phenomenon of green and tourism. The Lodge at Wah, Palampur is a perfect blend of green tourism for sustainable practices with production of organic tea. The initiatives provide transformative aspects with respect to developing green human resource with capacity building of local community which is established at the core of green tourism. The use of green architecture with construction of mud houses using slate and wood further create a sense of pride among local community with increase in interest of green tourist towards more eco-friendly staying options. Elshaer et al. (2023) assert farm to fork as a strategy to strengthen the local community while purchasing from farmers and make it a point of source for hotels and restaurants locally. Rakkh Resort Palampur provides a best example in production of local organic vegetables to be cooked in the kitchen to be served to the guests in the natural environment. The theoretical underpinning of green economy or business rests on principles of circular economy. Pan et al. (2018) advanced the thought of circular economy in green tourism highlighting firstly efficiency in resource management and secondly sourcing of local supplies, material and lastly services.

The transformative nature of support mechanism in promotion of green tourism in the Himalayan region comes from some specific out of box experiments in the field of sustainable development and its measurement. One of the first initiatives in the direction to protect environment and ecology with an aim to promote green tourism has been implementation of geographical indication tags to the products in the Himalayan region. Hazarhun and Tepeci (2018) highlighted the importance of geographical indications as at world level it has been realized that preserving and transference of local knowledge to the next generation can only be achieved through protection of indigenous knowledge and identified by the geographical indication (GI) tags. In the initial interventions with GI tags in the Himalayan region has been towards the identification of standing of Kangra Tea at the world level and green sustainable practices followed in line with the green tourism, i.e. circular economy, low carbon footprint, ecotourism, etc.

Other specific example to promote green tourism has been in the memorabilia production, giving boost to the initiative of green tourism. Memorabilia forms an important part of tourism industry where most of the tourist visiting the destination looks to take back memories in the form of souvenir. Green livelihood has been at the centre goal of green tourism. The same has been achieved with promotion of souvenir making in the Himalayan region with one of the most abundant tree waste, i.e. Pine Needle. The same has been used to guide and impart women for knowledge transfer with a focus on community tourism approach which is sustainable in nature and engage the free time of women in removing socio-economic disparities in the region.

Food has been recognized as an important parameter of tourism product creation (Jones & Jenkins, 2003) in an approach to organise the efforts of the local community to promote and patronise the local food items at the destination and at the host places too. Beer et al. (2003) do admit the acknowledgement of contribution of local food systems and provided viable examples of local gastronomy which further help in promotion of local economies, society at local level, culture and environment. Imbibing the knowledge dimension of the past in the Himalayan region promoted the art of pattal making (a utensil to eat the meal). The engagement of the local economies in collection of raw material from the forest, using the collected material for production of eating plates and serving of local meals items on it which is highly appreciated by tourists during their meal selection at the destination. In India, for example in the place 'queen of Hills' Shimla an authentic gastronomy experience is created with the establishment of eatery with the name 'Himachali Rasoi'. This initiative by the start-up has proven to be successful in attracting tourists to the destination and satisfying their taste buds with local cuisine. The activity is a link establishment with bonding to the destination, promotion of green tourism and revisiting the gastronomical history.

3. Discussion of the Practical Implications of Identified Transformative Aspects for Sustainable Tourism Development in the Himalayas

In recent times, extensive research has been conducted to understand the impact of external factors such as climate change, disease outbreaks and pandemics on consumer behaviour towards the environment. Today, individuals are inclined to travel with a focus on health safety and lifestyle enhancements, seeking balance within themselves and their surroundings. This change towards eco-friendlier tourist practices is seen by many as not just a passing fad, but rather a long-term and sustainable way of life (Mishulina, 2022).

Many people travel to the Himalayas in search of spiritual growth and restoration. Yet, the rise in tourism has brought about notable transformations including urbanisation, better infrastructure and more job prospects. On the other hand, this growth has also caused damage to the environment and disrupted traditional cultural practices along with it is imperative to acknowledge the harmful effects the hospitality industry has had on the Himalayan ecosystem, contributing to climate change and ecological harm. An increasing frequency of cloud bursts in arid highlands like Ladakh, and more recently in Kullu, Manali, Mandi, Shimla and Dharamshala due to the surge in tourist arrivals has exacerbated environmental issues in the region. Famous tourist destinations like Shimla and Manali in Himachal Pradesh and Nainital and Mussoorie in Uttarakhand have undergone significant changes in their ecosystems and are now facing the negative impacts of climate change. The region is experiencing glacier outbursts, cloud bursts, flash floods, heavy rainfall and landslides more frequently, which is affecting their natural habitat (Thayyen, 2020). The problems of vehicle pollution, traffic congestion, insufficient parking space, litter and improper waste management have become severe. These issues have strained the relationship between the locals and the visitors, and they have also caused damage to the fragile nature of the Himalayan region. The tourism activities are having a significant impact on the ecosystem and climate of the local community, which is a cause for concern. It is imperative to alter the tourism practices to protect the environment (Bhalla & Chowdhary, 2022).

Practical implications identified through transformative measures towards promotion of green tourism have allowed managing and multiplying the efforts in direction towards advancement of goal attainment under green destinations. In a nutshell it could be elaborated that green initiatives in the Himalayan region is a dedicated approach towards the loud voices in protection of ecology and maintain ecological balance. In the humble approach towards green tourism the Tourism Policy of Himachal Pradesh has taken substantive efforts like 'Green Protocol' with key highlights to reduce, reuse and recycle. The aim is to promote the feeling of self-destinations among the visitors with more awareness towards promoting green practices. The action-oriented approach in the Himalayan region comes from activities like solid waste management at all the destinations, water conservation, technology change through financial incentive, tourist buses using cleaner fuel, green tax from the vehicles entering major hot spots of tourism, eco labelling of tourism products and services and last but not the least making environmental audit to be mandatory exercise for all the tourism units. Capacity building has been at the centre focus of the Government of Himachal Pradesh to encourage green practices among the staff. The aim of transformative efforts in the state are highlighted in government initiatives which focus on green alternatives for the state like daylight/movement sensors in passageways, highest efforts towards solar energy usage, solar powered streetlights, rainwater harvesting and water saving fixtures, etc. The state of Uttarakhand in the Himalayan region focuses on low impact tourism with high encouragement to Greenfield infrastructure with close adaptation to the policy implications towards Indian Green Building Council guidelines. The mechanism to evaluate different tourism ventures comes from approach of green certification which provides a distinct identity to the organisation with critical measurement of parameters like energy efficiency, climate sound, water consumption, waste management, biodiversity conservation, preservation of cultural heritage, cultural understanding and tolerance, green building design and community collaboration.

The inventiveness of similar nature is well accepted at policy planner's level and implementing agencies which focus on transformational change to adaptation of local communities, stakeholders and tourism practitioners. The top-bottom inclusive approach allows to create an ecosystem based on the pillars of informed dialogue and largely based on evidence for public policy to be rational. It further leads to decision making which is highly based on the acceptance to principles of green tourism for effective decision-making for correctly estimating the expected benefits arising out of the same. Impact of the activity in the same direction has been more practical than thoughtful in green tourism. Local communities are mobilised towards the protection of flora and fauna with efforts introduced in the form of garbage segregation and disposal at the community level or local level. The examples come from respect towards environment and its protection. The best suited example in close intervention with the efforts are from AIMA Panchayat (a local village level organisation in India) in Palampur, Himachal Pradesh, Himalayan Region where local waste has been fruitfully transformed into useful waste using improvised local knowledge and experience. Further, opening of Tulip Garden in the Hills of Himalayas at city of Palampur is great experience for tourists and the local visitors to engage with colour therapy for improvement in health from environment initiatives. The last but not the least tourism industry engagement has been a key focus area in the Himalayan region as a major earner for the states with low per capita income and limited opportunities to earn from other activities like development of Industrial hubs for production of goods. The responsible behaviour in sustainable practices for green tourism can be well attributed to the efforts of ITC hotels towards green sustainability. ITC is one of the leading hotel chain in the world to covert its mechanism towards green sustainability like single use plastic, recycling of waste water, green infrastructure, etc.

4. Conclusion

Travel and tourism has become a popular avenue for personal growth and development. Among the most sought-after destinations for those seeking transformation are the Himalayan regions. The allure of these remote areas lies in the opportunity to escape the hustle and bustle of urban life and discover inner peace and tranquility amidst nature's beauty. Travelers are able to reflect on their personal and professional lives, making this journey a transformative experience – not just in a physical sense, but inwardly as well (Gills, 2020). Gill et al. (2019) stressed the significance of rejuvenating one's spiritual being, whereas Sheldon (2020) posited that travel can foster inner growth and a broadened mindset. The research indicates that mental health concerns can inspire a quest for holistic transformation and wellness, and those who undergo a shift in perspective and develop a deeper connection with nature are more inclined to seek out activities and events that promote restoration and purpose. Those with a heightened awareness value and respect all living things and the natural world. According to Ross (2010), transformation begins with a disorienting challenge that prompts a shift in both our mindset and behaviour. To fully leverage the advantages of regenerative tourism, it is crucial to prioritise personal growth and well-being. On the one hand, transformative tourism can contribute to the betterment of humanity, while regenerative tourism can promote environmental sustainability and ecological preservation.

Tourism must transform now more than ever. Its impact on climate change, reliance on neoliberal economics and struggle for sustainability in a world of over-tourism and over-consumption require a new approach. This shift in paradigm can only happen if everyone involved in tourism undergoes a transformation of their values and consciousness (Sheldon, 2020). The idea of human superiority over nature has led to the exploitation of resources and disregard for nature and living beings. The Covid-19 pandemic has given us a chance to reflect on our actions and allowed nature to heal. We need to awaken to a higher level of consciousness and transformation for the sake of people and the planet. This will establish a future where coexisting with nature is of utmost importance (Vuta, 2023).

The importance of transforming the tourism industry is highlighted by Pollock's observation that 'we need to change our perspective and align it with what both our hearts and modern science tell us' (Pollock, 2015, p. 6). As more travelers undergo personal transformations, tourism can become a driving force for positive change in society (Ateljevic, 2011). By embracing a regenerative economy that benefits all stakeholders, tourism can help create a more sustainable and fulfiling future (Pollock, 2017). Tourists and destination stakeholders who are responsible and aware have the potential to shape a new future for tourism. This future involves restoring and renewing our destinations and ourselves (Sheldon, 2020).

5. Future Research Direction

Based on the fruitful insights on the topic of green tourism, transformation, sustainable tourism and its implications at multiple levels a critical overview of the activities and initiatives undertaken is necessary at current times for future research direction. The recommendations for green tourism come from application of collaborative forecasting technique at inter and intra organisational level which allows operationalising the plan of action among various partner firms engaged in green economy. Secondly, using of the cluster data technique can provide major highlights to the efforts towards green tourism. Dolnicar (2002) do provide evidences that cluster data is a useful tool both for the researcher and the management for testing and validating the usefulness of collected data. Thirdly, a sector-oriented tool kit is must to adapt, accept and perform activities which are in support to green tourism. Fourthly, usefulness of computation of results from micro aggregate data of the visitors at the destination level highlights the key factors and reasons to select a destination. Individual micro data provides clear reasons for the choice of destination, the amount of spending and specific goods expenditure at the destination (Marrocu et al., 2015).

The topic of strategic communication for sustainable development sheds the light for next best possible solution in view to provide greater opportunity firstly to local community and secondly transform web of experience of the tourists through green tourism. Strategic communication for sustainable development allows putting people at the centre stage carrying out two-way transformative communication process while attaining higher multi order goals as inscribed in sustainable livelihood approach. The action-based approach involves people for dialogue, addressing the local needs with engagement, attention to sustainable development initiative and cohesiveness among members through collaborative action approach and alliances, bottom of the pyramid approach for mitigating the efforts of overall sustainable development (Chan et al., 2021). The entire process allows joint action in promotion of transformative aspects to build up an effective mechanism for transformative green tourism.

References

Al-Aomar, R., & Hussain, M. (2017). An assessment of green practices in a hotel supply chain: A study of UAE hotels. *Journal of Hospitality and Tourism Management*, *32*, 71–81. https://doi.org/10.1016/j.jhtm.2017.04.002

Allen, J. C., & Malin, S. (2008). Green entrepreneurship: A method for managing natural resources? *Society & Natural Resources*, *21*(9), 828–844. https://doi.org/10.1080/08941920701612917

Arnould, E., & Price, L. (1993). River magic: Extraordinary experience and the extended service encounter. *Journal of Consumer Research*, *20*(1), 24–45. http://www.jstor.org/stable/2489198

Ateljevic, I. (2011). Transmodern critical tourism studies: A call for hope and transformation. *Revista Turismo em Analise*, *22*(3), 497–515.

Ateljevic, I. (2020). Transforming the (tourism) world for good and (re) generating the potential 'new normal'. *Tourism Geographies*, *22*(3), 467–475. https://doi.org/10.1080/14616688.2020.1759134

Beer, S., Edwards, J., Fernandes, C., & Sampaio, F. (2003). Regional food cultures: Integral to the rural tourism product? In *Tourism and gastronomy* (pp. 221–237). Routledge.

Bhalla, R., & Chowdhary, N. (2022). Green workers of Himalayas: Evidence of transformation induced regeneration. *Journal of Tourism Futures*, *8*(3), 380–392. https://doi.org/10.1108/JTF-12-2021-0273

Bhatia, A. (2022). Determinants of ICT adoption for digital inclusiveness in hill state of Himachal Pradesh-tourists perspective. *Atna Journal of Tourism Studies*, *17*(1), 1–31. https://doi.org/10.12727/ajts.27.1

Borel-Saladin, J. M., & Turok, I. N. (2013). The green economy: Incremental change or transformation? *Environmental Policy and Governance*, *23*(4), 209–220. http://hdl.handle.net/20.500.11910/2860

Boswijk, A., Peelen, E., & Olthof, S. (2013). *Economy of experiences*. European Centre for the Experience Economy.

Buhalis, D. (2022). Tourism management and marketing in transformation: Preface. In D. Buhalis (Ed.), *Encyclopedia of tourism management and marketing*. Edward Elgar Publishing. (Forthcoming). https://www.academia.edu/44983865/

Cave, J., & Dredge, D. (2020). Regenerative tourism needs diverse economic practices. *Tourism Geographies*, *22*, 1–11. https://doi.org/10.1080/14616688.2020.1768434

Chan, J. K. L., Marzuki, K. M., & Mohtar, T. M. (2021). Local community participation and responsible tourism practices in ecotourism destination: A case of lower Kinabatangan, Sabah. *Sustainability*, *13*(23). https://doi.org/10.3390/su132313302

Coghlan, A., & Weiler, B. (2015). Examining transformative processes in volunteer tourism. *Current Issues in Tourism*. https://doi.org/10. 1080/13683500.2015.1102209

Dolnicar, S. (2002). A review of data-driven market segmentation in tourism. *Journal of Travel & Tourism Marketing*, *12*(1), 1–22. https://doi.org/10.1300/J073v12n01_01

Duxbury, N., Bakas, F. E., Vinagre de Castro, T., & Silva, S. (2020). Creative tourism development models towards sustainable and regenerative tourism. *Sustainability*, *13*(1), 1–17. https://doi.org/10.3390/su13010002

Elshaer, I. A., Azazz, A. M., Hassan, S. S., & Fayyad, S. (2023). Farm-to-fork and sustainable agriculture practices: Perceived economic benefit as a moderator and environmental sustainability as a mediator. *Sustainability*, *15*(14). https://doi.org/10.3390/su151411462

Furqan, A., Som, A. P. M., & Hussin, R. (2010). Promoting green tourism for future sustainability. *Theoretical and Empirical Researches in Urban Management*, *5*(8(17)), 64–74.

Gill, C., Packer, J., & Ballantyne, R. (2019). Spiritual retreats as a restorative destination: Design factors facilitating restorative outcomes. *Annals of Tourism Research*, *79*. https://doi.org/10.1016/j.annals.2019.102761

Gills, B. (2020). Deep restoration: From the great implosion to the great awakening. *Globalizations*, *17*(4), 577–579. https://doi.org/10.1080/14747731.2020.1748364

Hall, C., Scott, D., & Gössling, S. (2020). Pandemics, transformations and tourism: Be careful what you wish for. *Tourism Geographies*, *22*(1). https://doi.org/10.1080/14616688.2020.1759131

Han, H. (2020). Theory of green purchase behavior (TGPB): A new theory for sustainable consumption of green hotel and green restaurant products. *Business Strategy and the Environment*, *29*(6), 2815–2828. https://dor.org/10.1002/bse.2545

Han, H., Lee, K.-S., Radic, A., Ngah, A. H., & Kim, J. J. (2021). The extended self-identity-based electric product adoption model and airline business strategy: A new theoretical framework for green technology products. *Journal of Travel & Tourism Marketing*, *38*(3), 247–262. https://doi.org/10.1080/10548408.2021.1906386

Hazarhun, E., & Tepeci, M. (2018). The contribution of the local products and meals which have geographical indication to the development of gastronomy tourism in manisa. *Güncel Turizm Araştırmaları Dergisi*, *2*(Ek1), 371–389.

Higham, J., & Miller, G. (2018). Transforming societies and transforming tourism: Sustainable tourism in times of change. *Journal of Sustainable Tourism*, *26*(1), 1–8. https://doi.org/10.1080/09669582.2018.1407519

Horner, S., & Swarbrooke, J. (2020). *Consumer behaviour in tourism*. Routledge.

Jones, A., & Jenkins, I. (2003). 'A taste of Wales–Blas ArGymru': Institutional malaise inpromoting Welsh food tourism products. In *Tourism and gastronomy* (pp. 129–145). Routledge.

Kirillova, K., Lehto, X., & Cai, L. (2017a). What triggers transformative tourism experiences? *Tourism Recreation Research*, *42*(4), 498–511. https://doi.org/10.1080/02508281.2017.1342349

Kirillova, K., Lehto, X., & Cai, L. (2017b). Tourism and existential transformation: An empirical investigation. *Journal of Travel Research*, *56*(5), 638–650. https://doi.org/10.1177/0047287516650277

Kottler, J. A. (1998). Transformative travel. *Futurist*, *32*(3), 24–29.

Lean, G. L. (2012). Transformative travel: A mobilities perspective. *Tourist Studies*, *12*(2), 151–172. https://doi.org/10.1177/1468797612454624

Levin, J., & Steele, L. (2005). The transcendent experience: Conceptual, theoretical, and epidemiologic perspectives. *Explore: The Journal of Science and Healing*, *1*(2), 89–101. https://doi.org/10.1016/j.explore.2004.12.002

Marrocu, E., Paci, R., & Zara, A. (2015). Micro-economic determinants of tourist expenditure: A quantile regression approach. *Tourism Management*, *50*, 13–30. https://doi.org/10.1016/j.tourman.2015.01.006

Marsiglio, S. (2015). Economic growth and environment: Tourism as a trigger for green growth. *Tourism Economics*, *21*(1), 183–204. https://doi.org/10.5367/te.2014.0411

Mezirow, J. (1991). *Transformative dimensions of adult learning*. Jossey-Bass.

Ministry of Tourism. (2022, April 29). *National strategy for sustainable tourism*. Government of India.

Mishulina, S. I. (2022). Green transformation of the consumer behavior model of a tourism product. *Service in Russia and Abroad*, *16*(2/99), 24–39. https://doi.org/10.24412/1995-042X-2022-2-24-39

Morgan, A. D. (2010). Journeys into transformation: Travel to an "other" place as a vehicle for transformative learning. *Journal of Transformative Education, 8*(4), 246–268.

Pan, S. Y., Gao, M., Kim, H., Shah, K. J., Pei, S. L., & Chiang, P. C. (2018). Advances and challenges in sustainable tourism toward a green economy. *Science of the Total Environment, 635*, 452–469. https://doi.org/10.1016/j.scitotenv.2018.04.134

Pollock, A. (2015). Social entrepreneurship in tourism: The conscious travel approach. In *Tourism, innovation partnership for social entrepreneurship*. TIPSE Tourism.

Pollock, A. (2017, March 28). Sustainability is out and transformation is in–Beware. *Conscious. Travel Blog*. https://www.conscious.travel/sustainability-is-out-and-transformation-is-in-beware/. Accessed on 19 December 2019.

Prihanto, J. J. N., & Kurniasari, F. (2019, October). Sustainable digital transformation in the hospitality industry: A study of the hotel industry in Indonesia. In *2019 International Conference on Organizational Innovation (ICOI 2019)* (pp. 217–222). Atlantis Press. https://doi.org/10.2991/icoi-19.2019.38

Reisinger, Y. (2013). *Transformational tourism: Tourist perspectives*. CABI.

Ross, S. (2010). Transformative travel: An enjoyable way to foster radical change. *ReVision, 32*, 54–61. http://doi.org/10.4298/REVN.32.1.54-62

Sangiorgi, D. (2011). Transformative services and transformation design. *International Journal of Design, 5*(2), 29–40.

Schofield, P., Coromina, L., Camprubi, R., & Kim, S. (2020). An analysis of first-time and repeat-visitor destination images through the prism of the three-factor theory of consumer satisfaction. *Journal of Destination Marketing & Management, 17*, 100463. https://doi.org/10.1016/j.jdmm.2020.100463

Sheldon, P. J. (2020). Designing tourism experiences for inner transformation. *Annals of Tourism Research, 83*, 102935. https://doi.org/10.1016/j.annals.2020.102935

Siyambalapitiya, J., Zhang, X., & Liu, X. (2018). Green human resource management: A proposed model in the context of Sri Lanka's tourism industry. *Journal of Cleaner Production, 201*, 542–555. https://doi.org/10.1016/j.jclepro.2018.07.305

Snyder, C. (2008). Grabbing hold of moving target: Identifying and measuring the transformative learning process. *Journal of Transformative Education, 6*(3), 159–181. https://doi.org/10.1177/1541344608327813

Teoh, M. W., Wang, Y., & Kwek, A. (2021). Conceptualising co-created transformative tourism experiences: A systematic narrative review. *Journal of Hospitality and Tourism Management, 47*, 176–189. https://doi.org/10.1016/j.jhtm.2021.03.013

Thayyen, R. J. (2020). Hydrology of the cold-arid Himalaya. In A. Dimri, B. Bookhagen, M. Stoffel, & T. Yasunari (Eds.), *Himalayan weather and climate and their impact on the environment*. Springer. https://doi.org/10.1007/978-3-030-29684-1_20

Tussyadiah, I. P. (2014). Toward a theoretical foundation for experience design in tourism. *Journal of Travel Research, 53*(5), 543–564. https://doi.org/10.1177/0047287513513172

Voigt, C., Brown, G., & Howat, G. (2011). Wellness tourists: In search of transformation. *Tourism Review, 66*(1/2), 16–30. https://doi.org/10.1108/16605371111127206

Vuta, D. (2023, January 4). Green transformation strategies: Building sustainable tourism for the future. *Performance Magazine*. https://www.performancemagazine.org/green-transformation-strategies-building-sustainable-tourism-for-the-future/

Wahl, D. C. (2019). Travel to tomorrow 1: How we travel will affect where we arrive. *The Medium*. https://medium.com/age-of-awareness/travel-to-tomorrow-i-how-we-travel-will-affect-where-we-arrive-a214efc8099

Walker, R. H., & Hanson, D. J. (1998). Green marketing and green places: A taxonomy for the destination marketer. *Journal of Marketing Management*, *14*(6), 623–639. https://doi.org/10.1362/026725798784867671

Wang, J., Wang, S., Xue, H., Wang, Y., & Li, J. (2018). Green image and consumers' word-of-mouth intention in the green hotel industry: The moderating effect of Millennials. *Journal of Cleaner Production*, *181*, 426–436. https://doi.org/10.1016/j.jclepro.2018.01.250

Williams, K., & Harvey, D. (2001). Transcendent experience in forest environments. *Journal of Environmental Psychology*, *21*(3), 249–260. https://doi.org/10.1006/jevp.2001.0204

United Nations Environment Programme and World Tourism Organization. (2012). *Tourism in the green economy – Background report, UNWTO, Madrid*. https://doi.org/10.18111/9789284414529

Yamazaki, M. (1996). Green tourism in Japan. *Newsletter-Food & Fertilizer Technology Center*, *114*, 2–3.

Chapter 5

Sustainability and Innovation in the Wine Tourism Industry: The Case of Bodegas Franco-Españolas

Javier Martínez-Falcó[1], *Eduardo Sánchez-García*[2], *Bartolomé Marco-Lajara*[3] *and Joaquín Martínez-Falcó*

University of Alicante, Spain

Abstract

A captivating journey through Spain's verdant vineyards is unveiled in this study, where the burgeoning realm of wine tourism is revolutionising the wine industry from within. At the heart of this evolution lies Bodegas Franco-Españolas, where the fusion of wine culture, gastronomy and authenticity is being explored as a transformative force for sustainable and innovative practices. Through the lens of qualitative research, the myriad ways in which wine tourism acts as a catalyst for economic, social and environmental enrichment within the winemaking domain are revealed. This exploration transcends academic boundaries, offering a narrative of transformation that provides actionable insights and strategic frameworks for industry leaders and policymakers. The essence of wine tourism is being embraced, charting a course towards a future where the wine industry not only thrives but does so sustainably and innovatively. A toast is raised to a new era of wine tourism, where each glass poured signifies a step towards a greener, more vibrant viticultural landscape.

Keywords: Wine tourism; sustainability; innovation; Bodegas Franco-Españolas; Spain; wine industry

[1]https://orcid.org/0000-0001-9004-5816
[2]https://orcid.org/0000-0002-8524-4906
[3]https://orcid.org/0000-0001-8811-9118

Sustainable Tourism, Part B, 85–103

Published under exclusive licence by Emerald Publishing Limited
doi:10.1108/978-1-83608-326-920241005

1. Introduction

In the last two decades, a transformation in tourism demand trends has been observed, with a growing inclination towards shorter and more frequent trips (Festa et al., 2020). This dynamic has led to the emergence of innovative tourism modalities, in which gastronomy, culture and authenticity acquire a prominent relevance (Santos et al., 2019). In this context, at the end of the 20th century, the concept of enotourism emerged, a tourism typology focused on making known the wine culture and traditions of wine-producing regions (Trigo & Silva, 2022).

The first academic research in the field of wine tourism was developed by academics from New World institutions during the 1990s (Hall, 1997). Fundamental works in this field include Wine Tourism Around the World (Hall et al., 2000) and Explore Wine Tourism (Getz, 2000), recognised for their innovative approach, as well as the pioneering contributions of Charters and Ali-Knight (2002), Carlsen (2004), Getz and Brown (2006) and Mitchell and Hall (2006), for laying the foundations for the academic study of this type of tourism. Currently, wine tourism is a field of study of growing importance, with various ramifications having arisen around its study. In this sense, according to Gómez et al. (2019), six thematic axes can be fundamentally identified: (1) sustainable territorial progress, (2) wine routes, (3) wine tourists' trends, (4) the wine experience in wineries, (5) wine events and festivals and (6) wine marketing and promotion strategies. In this context, the study contributes to the axis of sustainable territorial progress, since the research aims to analyse the effect of wine tourism on improving the economic, social and environmental sustainability of wineries, as well as to examine its role on the development of organisational innovations in wineries.

From an economic perspective, wine tourism has become an additional source of income for wineries, diversifying their distribution channels and mitigating their vulnerability to fluctuations in the wine market (Andrade-Suárez & Caamaño-Franco, 2020). Likewise, by attracting tourists seeking exclusive experiences, an intrinsic value is added to the products and services offered (Nave & do Paço, 2021). In the social sphere, wine tourism favours the integration and collaboration of the local community in tourism activities, consolidating social cohesion and strengthening the sense of cultural identity (Oltean & Gabor, 2022). Similarly, by offering visitors an immersion in winemaking culture and traditions, it contributes to the preservation and enhancement of the local cultural heritage (Serra-Cantallops et al., 2021). From an environmental perspective, wineries with an enotourism orientation tend to be more proactive in the sustainable management of their resources, a fact that is reflected in the adoption of eco-friendly agricultural practices and the promulgation of tourism that seeks to minimise their ecological footprint (Karagiannis & Metaxas, 2020). In terms of organisational innovation, wine tourism, by presenting the challenge of serving a heterogeneous public, encourages innovation in services, infrastructure and experiences provided, which leads to greater responsiveness to changes in the market and the ability to anticipate future trends (Sousa, 2019).

However, despite the capacity of wine tourism to improve the sustainability of wineries in its triple dimension and enhance organisational innovation, there are several shortcomings in the study of these relationships. First, the academic literature has mostly focused on examining the link between wine tourism and the economic dimension of sustainability, leaving the social and environmental axis in the background. Second, there is little literature that has tried to link wine tourism with sustainability and organisational innovation together, representing an opportunity to further advance scientific knowledge linked to the benefits derived from wine tourism. Third, New World countries have received greater academic attention in terms of wine tourism compared to Old World wine nations, highlighting the need for further research on wine tourism in the Old World. Fourth, there is a notable paucity of research that has addressed such relationships in a qualitative approach, and there is a need to examine revealing case studies on the subject matter under study. Fifth, studies on sustainable wine tourism are limited, which can be attributed to the novelty of the concept of sustainability in the field of wine tourism, as well as to the complexity of integrating aspects of agriculture, tourism, environmental management and economics. Nevertheless, the study of this topic is of great importance due to its potential to mitigate the environmental impact of wine production and associated tourism, adapt to the challenges of climate change, contribute to local economic development, preserve traditions and cultures linked to viticulture, as well as educate visitors on responsible environmental practices. Research in sustainable wine tourism is therefore essential to ensure a balance between the economic and cultural benefits of this activity and the conservation of natural resources and social welfare.

Sustainable wine tourism represents a form of tourism that integrates the wine experience with a perspective of sustainability, seeking to harmonise the appreciation of wine and its culture with responsibility towards the social, economic and environmental surroundings. This form of tourism is based on practices that favour the conservation of the wine-growing landscape and biodiversity, ensuring the efficient management of resources and minimising environmental impact. In this context, environmentally friendly cultivation techniques are adopted, such as organic or biodynamic agriculture, and the use of renewable energies and waste reduction systems are promoted to reduce the carbon footprint of wine production.

From a social perspective, sustainable wine tourism boosts local development, generating employment and economic opportunities in host communities, while valuing and preserving traditions and culture linked to winemaking. This approach thus fosters community inclusion and social cohesion, enabling residents to actively participate in and benefit from tourism. Economically, it contributes to the financial sustainability of wine regions, since by attracting tourists interested in sustainable and conscientious practices, it generates an economic flow that supports the long-term viability of wineries and vineyards. In addition, sustainable management practices can lead to greater efficiency and profitability in wine production. Similarly, sustainable wine tourism also plays a crucial role in educating and raising environmental awareness among visitors about the importance of sustainability in viticulture and tourism, since, by offering experiences that highlight the interconnection between wine production and its impact on the environment and society, it promotes greater appreciation and respect for the conservation of natural resources and social well-being. In short, this form of tourism not only enriches the visitor experience but also promotes a balanced and responsible approach to wine production and tourism, aligned with the principles of sustainability.

In order to address the proposed research objective, it should be noted that the study is contextualised in the Spanish wine tourism sector, since in the last two decades, there has been a gradual process of intensification of wine tourism activity in wineries, as well as the certification of wine routes, allowing the accreditation of quality wine tourism activity in Spanish territories with a wine tradition. These wine routes, which amounted to 34 in 2022, have made it possible to articulate Spanish tourism activity, as well as to increase tourism flows associated with the world of wine, generating an economic impact of over 90 million euros, taking into account only visits to wineries and wine museums to these routes (Martínez-Falcó et al., 2023). In particular, the study focuses on the analysis of the case of Bodegas Franco-Españolas, located in Logroño (La Rioja), by offering the opportunity to study a phenomenon in its natural context, considering the interactions and particularities that surround it, allowing an exhaustive and in-depth analysis of the specific phenomenon analysed, providing a detailed understanding of the link between wine tourism and sustainability and corporate innovation, as well as offering practical recommendations for wine managers to improve their wine tourism activity, providing concrete examples and recommendations based on real experience. Through the analysis of the case examined, the aim is therefore to respond to the following two research propositions: (1) wine tourism favours the economic, social and environmental sustainability of wineries and (2) wine tourism catalyses the development of organisational innovations in wineries.

In order to respond to the two research propositions raised, after this introduction, the methodology followed in this study is presented in Section 2, the results are presented in Section 3 and, finally, Section 4 identifies the main conclusions, as well as the limitations and future lines of research.

2. Methodology

The methodology adopted in this research is aligned with the methodological foundations of the case study. As indicated by Rashid et al. (2019), the distinctive feature of this methodology lies in the comprehensive analysis of a clearly delimited entity, referred to as a case, which may represent an individual, a nation, an organisational entity or even a collectively. In such research, the case is meticulously analysed within its natural environment, highlighting its intricate nature (Alpi & Evans, 2019).

2.1 Sample

The selection of the sample is essential in research based on case studies, since the determination of the sample has a direct impact on the conclusions derived from the study (Welch et al., 2020). In the present study, a single case was chosen with the objective of deepening and acquiring a more detailed understanding of the phenomenon in question (Varela et al., 2021). For the choice of the case, leading organisations in sustainability and innovation within the wine tourism sector were recognised. According to the specialised literature, these leading organisations are those that take the lead in minimising the environmental impact derived from their activity and implement product and process innovations in their operations, thus obtaining, in this way, recognition for their ecological and innovative efforts in relation to other entities in the same sector (Santa-Maria et al., 2021). As for the selection of the winery, despite the non-probabilistic nature of the sampling, the following three criteria were established: firstly, that it had a comprehensive sustainability plan as proof of its commitment to the development of the triple bottom line; secondly, that it had received at least one 'Best of Wine Tourism' award, granted by the Great Wine Capitals Global Network, given the excellence in wine tourism innovation that this implies; and thirdly, that it had wide recognition both nationally and internationally in terms of sustainability and innovation. After an initial screening, Bodegas Franco-Españolas was selected as it met the three pre-established requirements.

2.2 Data Collection

Due to the qualitative approach of the research, triangulation was used in order to strengthen the validity of the study and confirm the deductions obtained from it (Hancock et al., 2021). To implement this triangulation, three sources of information were used: (1) direct observation was used, which included a visit to the winery's facilities on 7 October 2023 and the corresponding interaction with the staff, (2) an in-depth interview was conducted with the director of wine tourism, events and public relations, Elena Pilo González, on 25 October 2023 and (3) documentation was accessed both internally (statements on sustainability, newsletters, etc.) and externally (press articles, institutional videos, etc.).

3. Data Analysis

The case study research was structured in three stages: (1) review of previous literature, (2) detailed evaluation of the case in question and (3) analysis of the information collected. Initially, a detailed review of the academic literature related to sustainability and innovation in the wine tourism sector was carried out. Subsequently, a detailed analysis of the case of Bodegas Franco-Españolas was carried out to determine its relevance in the context of this research. In the last stage, information was obtained through an in-depth interview. This began with a block of questions about the history and origins of the winery. As the interview progressed, the questions became more specific, addressing the innovative practices developed by the organisation in wine tourism, as well as the link between this activity and the three pillars of sustainability (economic, social and environmental). The interview, of a structured nature, lasted one hour and was recorded in its entirety for later transcription. During the visit to the winery, dialogue was also established with other members of the team, concluding the data collection when the additional information did not provide greater clarity to the phenomenon studied (Bell et al., 2022). After transcribing the interview, the interviewees were asked to review and validate the content. Finally, once the transcripts were confirmed, a table was made to synthesise the research findings and proceeded to their interpretation.

4. Results

To facilitate the understanding of the results obtained, this section is structured in two blocks: (1) history and terroir and (2) wine tourism, sustainability and innovation.

4.1 History and Terroir

Nestled on the banks of the Ebro, in the embrace of the historic city of Logroño, Bodegas Franco-Españolas stands as the custodian of La Rioja's oenological heritage. More than a physical entity, it represents a living legacy that encapsulates the winemaking essence of Rioja, Logroño and Spain. With more than a century of existence, its walls have witnessed harvests and dispatched wines that distil the soul of their terroir.

The foundation of Bodegas Franco-Españolas originated in the perspicacity of Fréderic Anglade, a visionary from Bordeaux, who in 1890 established an avant-garde winemaking enclave in Logroño. Faced with the calamity of phylloxera in Bordeaux, winegrowers looked to Rioja wines as a substitute, thus cementing the reputation of the 'fine wines of Rioja'. The contemporary railway infrastructure was essential, promoting the distribution and commercial boom in a Logroño in full transformation and progress, symbolised by the inauguration of the iron bridge and the birth of key complementary industries. In 1891, Bodegas Franco-Españolas hosted its inaugural harvest, erecting its first two warehouses and thus forging its legacy. Deviating from the native carbonic maceration, winemaking was inspired by Bordeaux methods. From those early days, icons such as Diamante and Estilo Borgoña were born, renamed in the 1950s as Rioja Bordón and renamed in 2017 as Bordón. In the wine crisis of the late 19th century, the winery innovated by introducing vines on American rootstock, a strategy that preserved Rioja viticulture (see Fig. 5.1).

Over the years, the winery consolidated its business structure, and after two decades of operations, the Board of Directors already reflected a fully Spanish identity, counting among its members preeminent figures such as the Count of Romanones. The 1920s represented a period of prosperity for the winery, marked by events such as the visit of His Majesty King Alfonso XIII in 1925 and the creation of the Control Board of the Rioja Designation of Origin, the first in Spain, thus consolidating the region's reputation as a producer of quality wines. The winery's international projection was strengthened with the end of

Fig. 5.1. Centenary Vine of Bodegas Franco-Españolas. *Source:* Photograph by the authors. Date taken: 7 October 2023.

Prohibition in the United States, at which time emblematic Franco-Spanish brands such as Diamante and Royal began to position themselves on the menus of prestigious New York hotels. However, the Spanish Civil War was an unexpected setback with the closing of the borders, affecting foreign trade (see Fig. 5.2).

However, during the Second World War, Rioja wine exports found a new horizon. The war affected France, diverting the attention of markets such as Switzerland, Norway and Holland towards Rioja wines, thus benefiting Bodegas Franco-Españolas. Ernest Hemingway's association with the winery in the 1950s boosted its presence in the United States. The year 1964 was distinguished by an exceptional vintage, the best of the century, consolidating the winery in the US market. Its wines, served at distinguished events, became emblems of Spanish winemaking excellence at the international level.

The winery faced challenges with the adverse harvests of 1971 and 1972, aggravated by the oil crisis, culminating in its acquisition by Rumasa, which meant its integration into one of the most prominent conglomerates of the time. This turnaround inaugurated a period of change in the Spanish wine sector. In 1984,

Fig. 5.2. Italian Tanks in the Surroundings of Bodegas Franco-Españolas. *Source:* Photograph by the authors. Date taken: 7 October 2023.

under the direction of Marcos Eguizábal, a businessman with vision and passion for winemaking, Bodegas Franco-Españolas embarked on a path of renewal and excellence. In this sense, the commemoration of its centenary in 1990 reaffirmed its legacy and vision for the future. In 2012, Bodegas Franco-Españolas undertook the reinvention of its icons Diamante and Rioja Bordón, reflecting its adaptability to modernity while maintaining its legacy. In 2013, it was awarded 'The Best of' award in Wine Tourism for its innovative tourism proposals. The celebration of its 125th anniversary in 2015 marked a milestone of introspection and projection. In 2016, Borja Eguizábal inherited the management, ensuring the continuity of the exalted wine lineage. The year 2017 saw the rebirth of Rioja Bordón as Bordón, under the slogan 'Unabashedly classic', symbolising the harmony between tradition and contemporaneity. With more than 40,000 visitors in 2022, the winery reaffirmed itself as a wine tourism and cultural destination, offering everything from tastings to outdoor cinema, consolidating itself as a bastion of Rioja leisure and culture and perpetuating the Eguizábal legacy of innovation and quality (see Fig. 5.3).

Fig. 5.3. Current Facilities of Bodegas Franco-Españolas. *Source:* Photograph by the authors. Date taken: 7 October 2023.

4.2 Wine Tourism, Sustainability and Innovation

Wine tourism, which fuses wine appreciation with tourism, has become a strategic pillar in the management of modern wineries. For Bodegas Franco-Españolas, this practice exceeds the economic contribution, becoming a means of communication and brand positioning. Elena Pilo González, in charge of wine tourism, emphasises its importance not only as a sales channel but also as a thermometer to 'take the direct pulse of our customers', vital for the product strategy.

Wine tourism in Franco-Españolas wineries operates as a business entity with autonomy, managing independently, even in the use of the spaces, which are rented by the wine tourism division to the winery. This structure strengthens the identity and fosters the growth of wine tourism. Since 1915, the winery has understood the multifunctional value of its facilities, using them for events. In particular, the wine tourism team plays a dual role: it manages events and develops its own cultural agenda. This cultural integration into the wine tourism experience expands the winery's influence, making it an epicentre of leisure and cultural education. The winery's location, described by Pilo as 'privileged', is also a significant competitive advantage. The uniqueness of its location is an asset that, according to the head of wine tourism, gives them an undisputed position in the market, free from competition in similar conditions. This geographical factor is crucial, since proximity to areas of tourist interest and ease of access are determining factors in the choice of a winery to visit. The influx of more than 40,000 visitors per year, counting only tourist visits and excluding events, cultural activities and commercial visits, is indicative of the magnitude and success of wine tourism at Bodegas Franco-Españolas.

In relation to social sustainability, the winery is not only committed to creating quality wine tourism experiences but has also taken an active role in promoting culture and supporting the community. The organisation of both corporate and social events is a manifestation of this philosophy, where the winery becomes a space for cultural encounters and celebrations. The winery's cultural agenda, which includes activities such as summer cinema, theatre, music festivals and themed events such as 'Hallowine', reflects a significant investment in the social and cultural fabric of the region. With an annual allocation of approximately 150,000 euros for its cultural agenda, the wine tourism area of Bodegas Franco-Españolas demonstrates its capacity and commitment to reinvest in cultural activity. This investment not only promotes wine culture and its enjoyment but is also aligned with Corporate Social Responsibility (CSR) practices, contributing to cultural accessibility for the general public and supporting charitable causes and groups in vulnerable situations, as Pilo González points out. The figure of Ernest Hemingway, with his literary and personal influence, has played a fundamental role in the history and international positioning of the winery. The friendly relationship and the impact that Hemingway had on the winery (Fig. 5.4), especially in the recovery of exports and the opening of doors in the US market, are evidence of the importance of cultural ties and their contribution to the social sustainability of the company. In addition, the winery has integrated CSR into its broader sustainability plan (in the social dimension). This integration reflects a holistic approach in which CSR is not an isolated entity but an intrinsic part of the company's sustainability strategy. CSR actions include collaborations with educational institutions and universities, as well as initiatives that align with the social dimension of the sustainability plan, such as support for non-profit organisations.

Environmental sustainability is a critical dimension in the operation of contemporary wineries, and Bodegas Franco-Españolas has integrated this concern into its day-to-day management and long-term vision. The implementation of a waste

Fig. 5.4. Ernest Hemingway Together With Members of the Board of Directors of Bodegas Franco-Españolas. *Source:* Photograph by the authors. Date taken: 7 October 2023.

management plan and an energy resource savings plan are clear examples of how the winery is proactively addressing environmental challenges. The adoption of renewable energy technologies, such as solar panels, not only reflects a commitment to reducing the carbon footprint but also demonstrates an investment in energy efficiency and self-sufficiency, as Pilo González points out.

The winery's environmental strategy also includes minimal waste policies, especially in the restaurant and hospitality areas. In addition, it promotes collaboration with local businesses to strengthen the region's economy and reduce the environmental impact associated with long-distance transportation, which reinforces overall sustainability and supports the local community. In terms of infrastructure, the winery is embarking on a renovation project that prioritises improved accessibility, facilitating access to the wine tourism experience for a wider audience. Finally, the winery plans to incorporate the sustainability narrative into the visit experience, educating customers about sustainable tourism and how they can minimise their negative impact while enjoying the destination (see Fig. 5.5). This initiative, as Pilo González points out, not only adds value to the customer experience but also promotes responsible practices among visitors, aligning with a broader approach to sustainable tourism.

The development of innovations in the wine tourism sector, meanwhile, represents a crucial aspect of maintaining competitiveness and meeting changing consumer expectations. Bodegas Franco-Españolas has addressed this need through a series of technological and collaborative initiatives that have

Fig. 5.5. Guided Tour at Bodegas Franco-Españolas. *Source:* Photograph by the authors. Date taken: 7 October 2023.

transformed its commercial approach and interaction with customers. In 2016, the winery initiated a partnership with JIG, a local technology provider specialising in developments linked to the world of wine. This partnership led to the development of an ad hoc booking engine and the online marketing of its tourism services in 2018. Despite initial reluctance in the face of the loss of control that online sales could entail, as Pilo highlights, this strategy resulted in 34% growth in the first year, demonstrating the effectiveness of adapting to new digital consumption modalities.

The adoption of a booking engine and digital marketing strategies has catalysed new avenues of growth for Bodegas Franco-Españolas. The accumulation and analysis of visitor data has optimised conversion in the online store and enabled personalised communication, strengthening the relationship with the clientele. Likewise, the implementation of a robust data centre has facilitated the segmentation and enrichment of the database, enabling the automation of interactions with consumers. Similarly, the implementation of digital tasting has revolutionised the customer experience, allowing the collection of information from all attendees, not just the purchaser of the experience, thus expanding the winery's relational reach. What's more, in an effort to benefit the local economy, the winery, Pilo González notes, is developing a policy to suggest local establishments to visitors prior to their arrival at the winery, seeking a positive community economic impact. In addition, Bodegas Franco-Españolas has embraced sector collaboration, participating in Temporary Joint Ventures (UTE) to address shared challenges and promote innovative projects before the government of La Rioja. This collaborative approach not only benefits the winery but also drives the dissemination of innovative technology and practices throughout the wine industry (see Table 5.1).

Table 5.1. Content Analysis of the In-depth Interview.

Wine tourism	'Wine tourism, beyond its impact on the income statement, is a very interesting communication and brand positioning tool for us. And it also allows us to take the direct pulse of our customers. With what we get out of it, we make decisions regarding our main product, which is wine'.
	'The wine tourism area within the winery is a company. We are a company within the group and we act as such. In fact, we rent the facilities from Bodegas Franco-Españolas, although the owners are the same, which is the Eguizábal family. This gives us an entity as a business and, therefore, if we want this business to grow, we have to look at all its possibilities'.
	'On the one hand, our events team exploits the winery spaces for third parties and, on the other hand, we organise cultural events, a cultural agenda that does not come from the marketing department, surprisingly it comes from the wine tourism department'.
	'Now with the new project we have we are going to incorporate a wine bar [...] it is a way to retain the wine tourists themselves when they make the visit, but it has another approach to that tourist who does not want to visit the winery, many tourists have no interest in visiting the winery, they have already seen two, they do not want to see a third, what they want is to be able to taste some wines, to discover you and to be able to learn more about wine, but around wine they do not need the visit part'.
Economic sustainability	'In the winery there are a total of 50 workers and in wine tourism there are nine, without taking into account all the indirect hires. I Am talking about the Franco-Españolas staff. And we represent 10% of the profit and loss account'.

(Continued)

Table 5.1. *(Continued)*

	'The location we have is privileged and that gives us an undeniable competitive advantage. We are not going to have competition, of course, in the same conditions as us'. 'Today we receive more than 40,000 people in tourist visits alone. I am not taking into account events, cultural activities or commercial visits. Tourist visits alone, more than 40,000, to give you an idea of the scale'. 'The fact that we are an independent business unit gives us independence, redundancy aside, and at the end of the day we have a positive profit and loss account, which allows us to invest in all these activities'.
Social sustainability	'We do both corporate events and social events. Then there is the part of those events like the summer cinema, the diamond theatre, Halloween, which instead of Halloween is Hallowine, which we are doing now this weekend, a music festival called Mugui or Fashion Day. We have several events throughout the year that are what we call our cultural agenda'. 'More or less we allocate around 150,000 euros a year for the cultural agenda. The wine tourism area has, therefore, the capacity to reinvest 150,000 euros in cultural activity'. 'The wine tourism experience is linked to the figure of Hemingway. In the end, Hemingway had a great impact on this house. It's not just that Hemingway was here. It is thanks to him and thanks to the fact that he spoke about our wines in his books, in 56 when exports were recovered, that contact with Hemingway and that relationship of friendship that was generated with him opened doors for us in the North American market, which is still one of the best markets for us today'. 'We have a sustainability plan within which CSR is framed. In other words, the CSR part used to go it alone, but when we implemented a sustainability plan, CSR became part of the social dimension of the sustainability plan. For us, the entire cultural agenda is CSR. To give you an idea, not only because part of

Table 5.1. *(Continued)*

	the box office goes to non-profit organisations or groups at risk of exclusion or replanting forests when it's time, depending on the action we choose each year, but also because the entire cultural agenda is an investment for us. [...] But it is also CSR, for example the collaborations we do with schools, which we do a lot, or collaborations we do with the university or other series of actions that we incorporate and that we believe fit perfectly into the social dimension of our sustainability plan'.
Environmental sustainability	'Within the environmental dimension, we have a waste management plan and an energy resource saving management plan. Now the winery has incorporated the whole issue of solar panels and so on. And we get our supplies from the winery, so we are also, in a way, impacted by all of this. In addition, well, we have minimum waste policies in the whole of the restaurant and hotel sector'. 'And also, combining the social and environmental dimension, there are the policies of working with companies within our immediate environment'. 'We now have a reform project in which one of the things we are going to attack and which is a priority is to improve accessibility in terms of the winery's infrastructure'. 'One of the points we are going to tackle is that part in the storytelling of the visit to incorporate the whole part of sustainability, because we believe it adds value, but also to incorporate, as I said, communications to our customers about tourism sustainability and how they can improve or somehow reduce their negative impact when they are enjoying our destination'.
Organisational innovation	'In 2016 we started working with a technology provider, a partner called JIG, they are from Rioja and they have a lot of developments linked to the world of wine. And in 2016 we started to develop an ad hoc booking engine, and the online marketing of our tourism services in 2018. Initially, we were a little nervous at the beginning because selling online means not controlling what comes in. In the end, you put it into circulation, you market it, the bookings come in

(Continued)

Table 5.1. *(Continued)*

	and it's not like when you take them and manage them. Well, that online marketing made us grow by 34% in the first year'.
	'After implementing the booking engine and starting to sell everything online, we see not only other needs, but also a window of opportunity. With the data we capture from our visitors, we can try to improve conversion in the online shop. And not only in the online shop, we can start to communicate with our visitors in a personalised way and inform them of those activities that we are doing, both in the warehouse and elsewhere where we have our products. We have a tool called the CBD, which is a data centre. And through this data centre, what we do on the one hand is to have a very well-worked database, very well segmented, that we enrich the data as the user interacts with us and we activate a series of workflows. [...] As a result of this, other tools have been born. For example, digital tasting, which helps us to enrich the database. Think that in a reservation, generally, the tourist comes in a group. We capture the information about the person making the reservation, but not about the people accompanying them. So, through the digital tasting, we try to capture the data of all those who accompany them and continue communicating with these people'.
	'We are working on a policy whereby everyone who visits us, a few days before visiting us, is given a list of restaurants where they can go and a list of local shops where they can buy. With this, in a way, what we are trying to do is to give something back to the local population and to the society around us'.
	'We work a lot through associative projects with other wineries. We form a UTE, a temporary joint venture, and we work on an innovative challenge. We then present this challenge to the government of La Rioja to see if we can get funding and we usually get it. After that, it is easier for this technology to reach the rest of the wineries, because we have already made the initial effort'.

Source: Own elaboration.

5. Conclusions

Wine tourism is configured as a catalyst for economic development and a promoter of sustainability and innovation within the wine sector. Through the case study of Bodegas Franco-Españolas, it has been revealed that the strategic fusion of tourism with wine culture transcends the mere generation of income, exerting a positive impact on the socio-economic and ecological environment. Thus, the propositions put forward have been accepted, giving rise to an enriching debate on the theoretical, practical and political consequences of wine tourism for the industry and society as a whole.

In terms of theoretical implications, the research contributes to the academic literature by providing a detailed analysis of how wine tourism can act as a catalyst for comprehensive sustainability and innovation within the wine industry, particularly in the Spanish context. The study highlights the importance of adopting a strategic approach to the development of wine tourism, which can even become an independent unit of the winery. In this sense, to achieve such an approach, wineries can choose a central location, exploit the spaces of their facilities, develop an attractive cultural agenda, capitalise on the relationship with literary figures, develop corporate sustainability programmes, as well as digitise the tasting experience and booking activities. This allows driving sustainability and organisational innovation, as it fosters job creation around the wine tourism sector (economic dimension), catalyses the development of social programmes and the improvement of workers' welfare (social dimension), encourages the adoption of an environmental approach to attract wine tourists with a high ecological awareness (environmental dimension) and catalyses the development of innovations for the improvement of the wine tourism experience (organisational innovation).

From a practical perspective, the study shows that the adoption of digital technologies in wine tourism, specifically through digital tasting and reservation systems, enhances interaction with customers and optimises data management. This practice allows wineries to personalise communication and improve the visitor experience, which translates into greater loyalty and sales potential. Moreover, the implementation of marketing strategies based on data collection enriches the understanding of consumer behaviour and facilitates the adaptation of services to their expectations. At the policy level, the study highlights the importance of policies that encourage collaboration between wineries and other actors in the tourism sector. The formation of Temporary Joint Ventures (UTE) and the joint search for government funding for innovation projects are examples of how cooperation can accelerate the diffusion of technology and best practices in the industry. These alliances can be crucial for the sustainable development of wine tourism, aligning with policies that promote the local economy and environmental sustainability. Moreover, the recommendation of local establishments by public institutions to visitors reinforces the regional economic fabric and underlines the role of wine tourism as a catalyst for local economic development.

Despite the valuable insights offered by this study, it is acknowledged that the fact that it is based on a single case study introduces limitations in terms of the generalisability of the findings. This limitation stems from the specific nature of the winery analysed, which may not encapsulate the diverse spectrum of wine tourism practices prevalent in various regions. The selection of the Franco-Spanish winery was motivated by the intention to investigate in depth the specific relationships and dynamics within the field of wine tourism, this being a revealing case around the phenomena under study, which would allow for a richer understanding and a thorough exploration of the strategies employed. However, the study's focus on a single case inevitably restricts its ability to generalise conclusions and draw distinctions between different contexts, so, to overcome this limitation, it is proposed as a future line of research to conduct a multi-case analysis to establish comparisons around the relationships proposed. By adopting this comparative methodology, future research could contribute significantly to a more holistic and nuanced understanding of the interaction between wine tourism, innovation and sustainability across different wine contexts. Likewise, in order to be able to contrast whether the propositions put forward are positive and statistically significant in the population under study, the Spanish wine industry, it is proposed as a future line of research to carry out a quantitative analysis through structural equation modelling, proposing a theoretical model and, subsequently, contrasting it through the collection of primary data.

References

Alpi, K., & Evans, J. (2019). Distinguishing case study as a research method from case reports as a publication type. *Journal of the Medical Library Association*, *107*(1), 1.

Andrade-Suárez, M., & Caamaño-Franco, I. (2020). The relationship between industrial heritage, wine tourism, and sustainability: A case of local community perspective. *Sustainability*, *12*(18), 7453.

Bell, E., Bryman, A., & Harley, B. (2022). *Business research methods*. Oxford University Press.

Carlsen, P. (2004). A review of global wine tourism research. *Journal of Wine Research*, *15*(1), 5–13.

Charters, S., & Ali-Knight, J. (2002). Who is the wine tourist? *Tourism Management*, *23*(3), 311–319.

Festa, G., Shams, S., Metallo, G., & Cuomo, M. (2020). Opportunities and challenges in the contribution of wine routes to wine tourism in Italy–A stakeholders' perspective of development. *Tourism Management Perspectives*, *33*, 100585.

Getz, D. (2000). *Explore wine tourism: Management, development & destinations*. Cognizant Communication Corporation.

Getz, D., & Brown, G. (2006). Critical success factors for wine tourism regions: A demand analysis. *Tourism Management*, *27*(1), 146–158.

Gómez, M., Pratt, M., & Molina, A. (2019). Wine tourism research: A systematic review of 20 vintages from 1995 to 2014. *Current Issues in Tourism*, *22*(18), 2211–2249.

Hall, C. (1997). Geography, marketing and the selling of places. *Journal of Travel & Tourism Marketing*, *6*(3–4), 61–84.

Hall, C., Sharples, L., Cambourne, B., & Macionis, N. (2000). *Wine Tourism Around the World, Development management and markets.* Butterworth Heinemann.

Hancock, D., Algozzine, B., & Lim, J. (2021). *Doing case study research: A practical guide for beginning researchers.* Teachers College Press.

Karagiannis, D., & Metaxas, T. (2020). Sustainable wine tourism development: Case studies from the Greek region of Peloponnese. *Sustainability, 12*(12), 5223.

Martínez-Falcó, J., Marco-Lajara, B., Zaragoza Sáez, P., & Sánchez-García, E. (2023). Wine tourism in Spain: The economic impact derived from visits to wineries and museums on wine routes. *Investigaciones Turísticas, 25*, 168–195.

Mitchell, R., & Hall, C. M. (2006). Wine tourism research: The state of play. *Tourism Review International, 9*(4), 307–332.

Nave, A., & do Paço, A. (2021). Sustainability in the wine-tourism sector–an analysis of perceived understanding and practices implemented by firms. *Journal of Wine Research, 32*(2), 103–116.

Oltean, F., & Gabor, M. (2022). Wine tourism – A sustainable management tool for rural development and vineyards: Cross-cultural analysis of the consumer profile from Romania and moldova. *Agriculture, 12*(10), 1614.

Rashid, Y., Rashid, A., Warraich, M. A., Sabir, S., & Waseem, A. (2019). Case study method: A step-by-step guide for business researchers. *International Journal of Qualitative Methods, 18*, 1609406919862424.

Santa-Maria, T., Vermeulen, W., & Baumgartner, R. (2021). Framing and assessing the emergent field of business model innovation for the circular economy: A combined literature review and multiple case study approach. *Sustainable Production and Consumption, 26*, 872–891.

Santos, V. R., Ramos, P., Almeida, N., & Santos-Pavón, E. (2019). Wine and wine tourism experience: A theoretical and conceptual review. *Worldwide Hospitality and Tourism Themes, 11*(6), 718–730.

Serra-Cantallops, A., Ramón-Cardona, J., & Vachiano, M. (2021). Increasing sustainability through wine tourism in mass tourism destinations. The case of the Balearic Islands. *Sustainability, 13*(5), 2481.

Sousa, B. (2019). A theoretical contribution from the perspective of innovation process in wine tourism contexts. *Marketing & Tourism Review, 4*(2), 1–18.

Trigo, A., & Silva, P. (2022). Sustainable development directions for wine tourism in Douro wine region, Portugal *Sustainability, 14*(7), 3949.

Varela, M., Lopes, P., & Rodrigues, R. (2021). Rigour in the management case study method: A study on Master's dissertations. *Electronic Journal of Business Research Methods, 19*(1), 1–13.

Welch, C., Piekkari, R., Plakoyiannaki, E., & Paavilainen-Mäntymäki, E. (2020). Theorising from case studies: Towards a pluralist future for international business research. *Research Methods in International Business, 42*, 171–220.

Chapter 6

Reducing Carbon Footprints Through Green Technology in the Tourism Industry: A Review of Selected Hotels in Delhi

Rekha Maitra[1]

G.D. Goenka Public School, India

Abstract

The carbon footprint measures a company's contribution to climate change. Implementing guidelines to address the rising carbon footprint while maintaining environmental compliance is essential. This chapter aims to identify existing sustainable practices, their benefits and drawbacks and examine the gaps in reducing the carbon footprint. Managing carbon footprint in sustainable tourism development is crucial today, where tourism has become a billion-dollar industry. Very few studies have investigated barriers to carbon footprint reduction programmes. This paper discusses the importance of reducing carbon footprints in the tourism industry. The five research objectives are (1) To judge the key factors contributing to the tourism industry's carbon footprint. (2) How can sustainable tourism development help reduce the carbon footprint? (3) What practical steps can tourism businesses take to manage their carbon footprint? (4) What is the impact of digital transformation on environmental sustainability? And (5) To identify the scope of an interdisciplinary approach to sustainability and technology transformation. To this end, the researcher has chosen the literature mapping method to examine the impact of digital transformation on environmental sustainability and identify the scope of an interdisciplinary approach to sustainability and technology transformation. Mapping has been described variously as a 'graphic blueprint', a 'diagrammatic representation' and a 'geographical metaphor' of the research field. Maps provide 'tangible

[1]https://orcid.org/0000-0001-9293-1969

Sustainable Tourism, Part B, 105–121

Published under exclusive licence by Emerald Publishing Limited
doi:10.1108/978-1-83608-326-920241006

evidence' of a student's comprehension and interpretation of the research domain that can be shared with peers and supervisors.

Keywords: Net zero; sustainability; responsibility; sensitisation; technology

1. Introduction

Carbon footprint is a widely used measure of an organisation's contribution to climate change. Our actions' total greenhouse gases (including carbon dioxide and methane) are called our carbon footprint. Food, consumption, transportation, and household energy have the highest carbon footprints. Hotels contribute significantly to environmental degradation, particularly regarding their carbon footprint. A hotel's carbon footprint includes all the greenhouse gas emissions produced during its operations, from energy consumption to transportation, waste disposal and more. Reducing a hotel's carbon footprint is crucial to ensuring environmental sustainability. Hotel operations consume significant energy and water, resulting in carbon emissions (Schubert et al., 2010).

One of the main drivers of a hotel's carbon footprint is energy consumption. Hotels use energy extensively for heating, cooling, lighting and other purposes. To reduce energy consumption, hotels can adopt energy-efficient technologies, such as LED lighting, innovative heating and cooling systems and energy-efficient appliances. Another way to reduce hotels' carbon footprint is to use renewable energy sources, such as solar or wind power.

The study by Chan (2021a) delves into the strategies employed by hotel managers to reduce their carbon footprint by engaging stakeholders. The findings reveal that hotel owners, employees, customers, governments, and others play crucial roles in these strategies, and the various influencing strategies used with these stakeholders are thoroughly discussed.

Singh et al. (2014), hotels also generate significant amounts of waste, which can contribute significantly to greenhouse gas emissions. Hotels can implement recycling programs, composting and source reduction strategies to reduce waste.

The carbon footprint of an individual, organisation, or product can be calculated by assessing the emissions associated with energy consumption, transportation, waste management and other activities. Reducing carbon footprint is crucial for mitigating climate change, and there are many ways to do so, such as using renewable energy sources, reducing energy consumption and adopting sustainable transportation options.

Bănescu et al. (2021), due to digitalisation, tourism has grown in profitability, competitiveness and sustainability. People have access to viewing unknown places. It has been observed that the tourism and hotel industries are interdependent; their existence will only be prevalent with each other. Customers buying the tour package will opt for transportation, accommodation and food. The buyer and seller need to follow sustainable practices, i.e. digitalisation.

Abeydeera and Karunasena's (2019) study on the carbon emissions of the Sri Lankan hotel industry reveals significant insights into the sector's environmental impact. The research aimed to explore the carbon emissions of hotels in Sri Lanka and to develop a standardized emission reporting guideline for the industry. The study underlines the importance of implementing effective carbon management practices in Sri Lankan hotels to mitigate their environmental impact and align with global sustainability goals. Dubois and Ceron (2006) emphasised that the importance of lowering sustainable house gas emissions has been signified by International researchers who are finding methods of reducing hotel carbon emissions. However, hotel carbon emissions remain an unexplored field in developing countries.

Hotels have installed energy savers in their rooms, washrooms, and public areas to optimise energy usage and have adopted strict environmental measures. To reduce energy consumption, but the sensitisation of hotel employees remains a concern due to the high turnover rate of hotel employees. Therefore, academic research is more critical to find new solutions. Digitalisation needed more presence in the 2030 Agenda or the Paris Agreement. It is evident that digital transformation is becoming a catalyst for social change. The digital alteration will significantly affect every aspect of global societies and economies. Digitalisation is not just an 'instrument' to resolve environmental difficulties; it also carries immense disruption potential.

2. Advantages of Carbon Emission Reduction in the Hotel Industry

TWI2050 – The World in 2050 (2019) report examined the potential effects of digitalisation on sustainable development for all. It draws attention to both positive outcomes and potential risks, particularly for the disadvantage of the 'digital divide'. The preconditions necessary for a successful digital transformation – such as prosperity, social inclusion, environmental sustainability and good governance – are explored in detail. It expresses the implications of digitalisation on society and addresses an area not covered sufficiently in discussions between sustainability and the Digital Revolution: governance elements. Fig. 6.1 above illustrates the geographical divide of global tourism over a 3 year period.

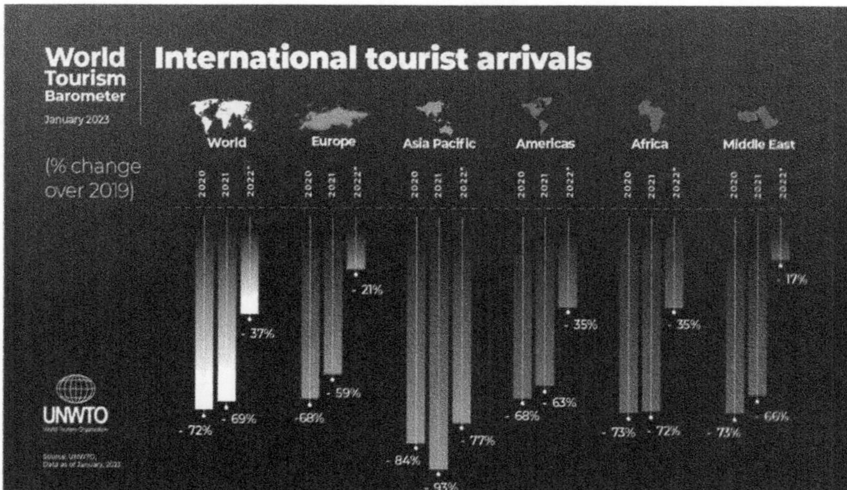

Fig. 6.1. World Tourism Barometer, 2022. *Source:* UNWTO (2022).

Sustainable Hospitality Alliance CEO Glenn Mandziuk emphasised recognising the importance of protecting our planet as an industry dependent on the natural world, from our buildings to attracting guests to outstanding locations around the world (UNWTO, 2022).

According to Thanh Cong Truong (2022), The Digital Revolution is a powerful catalyst for social inclusion. To ensure an effective transition towards sustainability, the 2030 Agenda goals must be attuned to the opportunities and risks that Digital Dynamics present. Digitisation will impact economies and cultures worldwide, altering approaches to attaining sustainable development to minimise adverse social, economic and environmental impacts.

The key point that needs to be focused on in the next era of sustainability is reducing carbon footprint are as shown in Fig. 6.2.

Technological innovations and systemic changes will promote renewable energy, electrification of transportation and circular economy practices. This may involve the widespread adoption of electric vehicles, the development of advanced renewable energy technologies such as solar power and hydrogen fuel cells, and the implementation of circular economy principles in industries such as fashion and agriculture. Additionally, there may be a growing emphasis on carbon capture and storage technologies and using nature-based solutions such as reforestation and soil carbon sequestration to mitigate climate change.

Customer engagement is crucial for the success of sustainable solutions. It involves educating and empowering customers to make informed decisions about their energy usage and encouraging them to adopt sustainable practices. A strong distribution network can help reach a wider audience and increase the adoption of sustainable solutions. Businesses can create a sustainable future for their

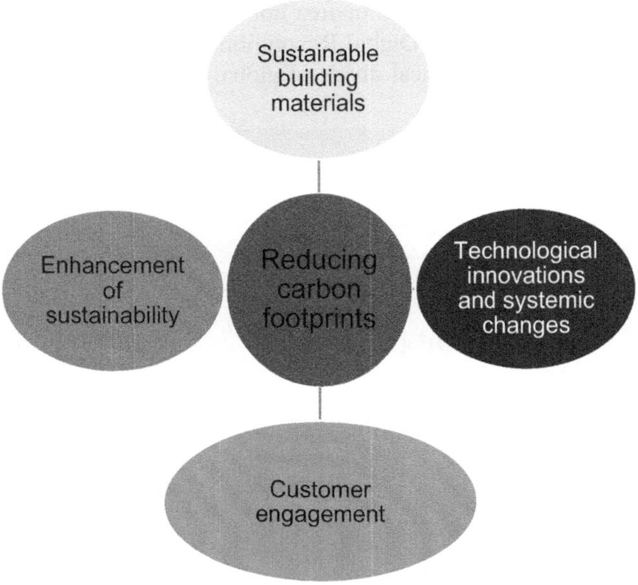

Fig. 6.2. Key Points to Reduce Carbon Footprint. *Source:* Author's work.

customers and the environment by engaging customers and leveraging a robust distribution network.

Enhancement of sustainability measures refers to improving or strengthening existing sustainability practices to achieve better environmental, social and economic outcomes. This can involve implementing new technologies, processes or policies that reduce waste and emissions, promote renewable energy and support sustainable development. By enhancing sustainability measures, organisations can reduce their environmental impact, improve their reputation and contribute to a more sustainable future.

Sustainable building materials play a crucial role in reducing carbon footprint by minimising the environmental impact of construction and operation of buildings. These materials are made from renewable resources, have lower embodied energy and can help reduce waste. Examples of sustainable building materials include recycled concrete aggregate, bamboo and low volatile organic compound (VOC) paints. By incorporating these materials into building design, architects and builders can significantly reduce the carbon footprint of a building over its lifecycle.

3. Literature Review

Chan (2021b) analysed the barriers to reducing a hotel's carbon footprint and how stakeholders can coordinate better to achieve this goal. Through qualitative research, interviews with 22 hotel executives revealed that understanding, project ownership, accurate measurement and risk investment were all influential factors in making progress. It was highlighted that a strong mediator should be appointed to balance interests successfully.

"An empirical study of China's 92 tourism-dependent cities" by Tong et al. (2022) found that tourism positively impacts reducing carbon emissions, with regional variations. This highlights its potential for promoting sustainable development and lowering carbon emission intensity. In the paper by Sun et al. (2020) titled "Managing Tourism Emissions through Optimizing the Tourism Demand Mix: Concept and Analysis", the authors note that the tourism Industry's response to climate action, such as engaging businesses in technology adoption and encouraging more sustainable visitor behaviour, has been insufficient to mitigate tourism's soaring carbon footprint.

With over 900 million international tourists travelling each year before the pandemic (UNWTO Barometer, 2022), the importance of effective carbon footprint reduction is evident. Hotel managers must actively encourage stakeholders and ensure they collaborate towards achieving this common goal. For an overview of the literature, see Table 6.1.

4. Discussion

4.1 Hotels Must Invest in Technology Solutions That Require Less Hardware

Contemporary guests are environmentally conscious; therefore, following environmental policies' responsibility is critical. A hotel app allows guests to check in and select the services needed during their stay. Hotels can reduce the use of

Table 6.1. Literature Review Analysis.

Sl No.	Authors	Topic	Findings
1	Al et al. (2019)	The digital revolution and sustainable development: opportunities and challenges	The chapter provides a comprehensive literature review on the intersection of digital technology and sustainable development, highlighting the importance of a multidisciplinary approach that balances the potential benefits and challenges of the digital revolution.
2.	Miola (2019)	The digital revolution and sustainable development: opportunities and challenges	Digital technologies can help achieve the sustainable development goals by improving access to healthcare, education and energy services. However, challenges like the digital divide, cybersecurity risks and impact on jobs and income inequality must be addressed. Artificial intelligence may support sustainable development, but we must consider its risks and ethical implications.
3.	Cheng et al. (2023)	'A digital transformation approach in hospitality and tourism research'	The findings of this paper highlight the significance of digital transformation by providing a

Table 6.1. *(Continued)*

Sl No.	Authors	Topic	Findings
			theoretical research framework for digital transformation in hospitality and tourism, including possible directions, contextual factors and methodologies.
4.	Singh et al. (2014). International journal of hospitality management, 43, 13–22	Green strategies for hotels: estimation of recycling benefits	The article 'Green Strategies for Hotels: Estimation of Recycling Benefits' by Singh et al. (2014) discusses the importance of recycling in the hospitality industry. The authors estimate the benefits of recycling in hotels, including reduced waste disposal costs and improved environmental sustainability. The study provides practical recommendations for hotels implementing recycling programs and reducing their environmental impact.
5.	Hu et al. (2015)	Assessing carbon footprint in the life cycle of accommodation services: the case of an international tourist hotel	Analysed greenhouse gas emissions from an international hotel in Taiwan and found that energy consumption is the main contributor to

(Continued)

Table 6.1. *(Continued)*

Sl No.	Authors	Topic	Findings
			the hotel's carbon footprint. Amenities production, transportation and laundry services exceeded the 5% cut-off rule set by PAS 2050. It's essential to consider these factors when evaluating carbon emissions. Hotel occupancy rate is also crucial to carbon emissions during one-night stays in standard rooms.

Source: Author's work.

paper, save tonnes of water and chemicals used in paper processing, and eliminate wasteful plastic key cards by utilising digital devices. When combined with modern property-management-system platforms, essential mobile software can help transform the guest journey while saving costs.

Sustainability will be shaping the future of travel. Together, we can continue to make it easier for everyone to experience the world mindfully and responsibly. To build a decarbonised and sustainable future for India, the tourism industry must address common environmental issues and seek advanced ecological solutions.

The rapid development of information technology has influenced tourism, hospitality and events. Mobile application bookings led the last decade, the Internet of Things (IoT), artificial intelligence (AI), big data, cloud computing, augmented reality, virtual reality, meta verse and web 4.0 applications. Searching with filters, options for flexible bookings and extended stays with AirBnB are paving the way for future generations.

Most of the technology platforms have emerged in the last two decades. This has streamlined the process with efficacy. Most businesses have adopted cloud computing to operate and sustain their income adequately. Maitra (2021) highlighted that 'the tourism and hotel industries have embraced digital transformation to maintain their business. Although business digitalisation has been implemented over the last decade, full acceptance of the digital revolution has been observed post-pandemic'.

4.2 Role of Artificial Intelligence

Artificial intelligence systems have the potential to establish the link between a business and a consumer by producing reliable information to understand consumer needs. Artificial intelligence capabilities may assist businesses in retaining customers by improving customer engagement. The Internet has revolutionised the tourism and hospitality industries. The internet journey started with Web 1.0, followed by Web 2.0, Web 3.0 and Web 4.0, which has led to the acceptance of technology-led environments, impacting the connectivity of travellers and tourists worldwide. Simultaneously, the tourism and hospitality businesses have gained more than ever. Consumers' behaviour may range from what others are doing, information search, purchase decisions and post-purchase action, vis-à-vis what other suppliers offer. Many studies reviewed and investigated the seamless flow of work on technology-integrated platforms for the increasing role of tourism development.

Because of the critical role of digitisation in reducing the carbon footprint in the hospitality and tourism industry, the growth in the research on tourism and technology integration has increased manifold. Recent studies report that information related to travel decision-making is searched on virtual platforms, vlogs and blogs. Therefore, business managers, planners and decision-makers are interested in understanding the impact of technological advancements on tourism and hospitality. Academia is also more interested in deciphering how future managers and planners are progressing in this direction.

4.3 Adoption of Technological Advantages by Hotels

New technological innovations have led to new experiences in the tourism and hospitality industries. Hotel apps allow guests to quickly check-in, select services, and stay safe while reducing paper usage and costs. Advanced technology is improving customised software, energy efficiency and reducing emissions.

The food and beverage tourism industry has embraced augmented reality (AR) technology to enhance customer experiences. AR-enabled food apps allow customers to preview the aroma of food before placing an order, creating a unique dining experience. Additionally, this technology provides real-time location and mobile travel guides, offering insights into food and beverage destinations.

According to Silva (2003), AR is a technology that utilises computer graphics, multimedia and animation to represent a virtual world. Jung (2014) believes that it will help intensify the accommodation industry by creating digital environments of cultural artefacts and increasing tourists' footfalls in food and beverage tourism marketing.

4.4 Novel Techniques to Reduce Carbon Emissions in the Hospitality and Tourism Industry

They are conferring to (Dodds, et al., 2012) the study on the status of the voluntary carbon offset market in North America and the tourism industries' work towards mitigating climate change impacts. The carbon offsetting market has developed globally due to the increasing concern around greenhouse gas emissions generated by government agencies, academics, non-governmental organizations (NGOs) and businesses. However, North America was largely stagnant in compliance and voluntary offsetting market, presented as a short-term solution for mitigating climate change. The lack of standards and the credibility of offsets have confused consumers concerning their impact on reducing GHG emissions.

Maitra, Shukla, et al. (2023) emphasised the relationship between social media and sustainable tourism. The authors analyse the impact of social media on sustainable tourism practices and consumer behaviour. They find that social media can positively influence sustainable tourism by promoting eco-friendly practices and responsible travel behaviours. The study highlights the potential of social media to promote sustainable tourism and encourage responsible travel behaviours.

The effects of carbon dioxide emissions on climate change are readily apparent and substantiated by many scientists. The escalation of CO_2 output due to energy requirements and fuel use risks our ability to be sustainable. Studies have uncovered that tourism may be a producer of CO_2, thus making it a major contributory factor towards global warming. Carbon dioxide levels emanating from the three main branches of tourism – transport, hospitality and attractions – have been measured and calculated using various methodologies. It is estimated that this industry causes 5% of total CO_2 worldwide. The core aims are to assess the divergence in CO_2 production among transport means, lodging institutions, attractions, activities, and events within tourism, plus examine how visitors' travel style can impact these emissions.

The study by Maitra, Shukla, et al. (2023) explores the potential of social media in promoting renewable energy usage. The authors argue that social media can play a crucial role in raising awareness about the benefits of renewable energy, mobilising public support and influencing policy decisions. They suggest that social media platforms can be leveraged to share information, promote renewable energy projects and encourage public participation in the transition to sustainable energy sources. The article 'Internet of Things as a Sustainable Energy Management Solution at Tourism Destinations in India' by Tiwari et al. (2022) discusses the potential of the IoT to manage energy sustainably in tourist destinations in India. The authors argue that IoT can help optimise energy consumption and reduce waste through intelligent energy management systems, which can be particularly beneficial in areas with limited energy resources. They suggest IoT can monitor and control energy usage in tourist facilities, such as hotels and restaurants, and provide real-time data on energy consumption to inform decision-making. For an overview of initiatives taken to reduce carbon footprint by hotels, see Table 6.2.

5. Findings/Results

(1) *Oberoi Hotels* in India have implemented various sustainable practices to reduce their carbon footprint. These include using solar power, implementing energy-efficient lighting and HVAC systems, and sourcing local and organic

Table 6.2. Comparison of Initiative Taken to Reduce Carbon Footprint by Hotels.

Sl No.	Name of the Hotel	Initiatives Taken to Reduce Carbon Footprint by Hotels						
1	Hotel Oberoi (East India Hotels Limited)	Green Team	Energy conservation and management	Land use planning and management	Minimising and recycling waste	Preservation of air quality	Replacing plastics	Water management
2	Hotel Lalit	implementing energy-efficient measures, such as LED lights, solar panels and efficient HVAC systems	Implemented waste reduction strategies, such as composting and recycling. Rainwater harvesting and reusing greywater to reduce water consumption	Guests to participate in eco-friendly practices during their stay	Reducing their impact on the environment and being a sustainable hospitality brand			
3	Indian Hotels Company (IHCL),	Utilising climate-responsive elements such as glass, insulation,	Energy-efficient technology for cooling, hot water,	Lighting such as solar hot water systems	Heat pumps and recovering waste heat	Using energy from a renewable resource	Increased utilization of green sources of	Driving responsible tourism with

(Continued)

Table 6.2. *(Continued)*

Sl No.	Name of the Hotel	Initiatives Taken to Reduce Carbon Footprint by Hotels						
	Taj Group of Hotels	ventilation and daylighting reduces air conditioning requirements by up to 30%			for hot water to reduce energy consumption by over 60%		energy to 93% from its earlier 27%	sustainable practices
4	ITC Hotels	Harnessing renewable energy	Recycling water	Treating solid waste	Sourcing consumables locally	Eco-sensitive architecture	LEED Platinum Certified	Water conservation
5	Radisson Blu	Supports sustainability objectives	A leading hospitality industry tool, the Hotel Carbon Measurement Initiative (HCMI), was developed to consistently measure and analyse the carbon footprint of hotel stays and meetings. The HCMI framework calculates the carbon emissions allocated to a meeting space to offset them accordingly					

Source: Author's work.

produce. The hotels also have recycling programs and reduced water consumption through low-flow fixtures. Additionally, Oberoi Hotels has introduced electric vehicle charging stations and promoted eco-friendly packaging materials (Limited, 2024).

(2) *LALIT hotels* in India have implemented various practices to reduce their carbon footprint and minimise their environmental impact. These include using solar power, implementing energy-efficient lighting and appliances and sourcing local and organic produce. LALIT hotels have introduced recycling programs and encourage guests to reduce plastic usage. By adopting these practices, hotels aim to lower their carbon footprint and contribute to a sustainable future. LALIT hotels have recycling programs. LALIT Hotels encourages guests to reduce plastic usage (https://www.thelalit.com/wp-content/uploads/2021/01/We-Care-Booklet-Final.pdf, 2022).

(3) *IHCL*, the Indian hospitality company, has a supply chain engagement policy that includes using resources responsibly, reducing the use of resources like energy, water, packaging and other materials and minimising carbon emissions and waste generated. IHCL continues to endeavour to reduce its carbon footprint, with 35% of its hotels in south India utilizing energy from renewable sources (2022). IHCL has also set sustainability goals for 2030, including:

- Reducing greenhouse emissions by about 232,000,000 kilograms of carbon dioxide equivalent (CO_2e).
- Reducing water consumption by about 3,000,000 kilolitres.
- Saving about 1,202,000,000 megajoules of energy.
- Reducing waste sent to landfills by 17,000 cubic metres.
- Deriving 50% of its energy across all hotels from renewable sources by 2030.
- Phasing out single-use plastics from all its hotels.
- Replacing plastic water bottles with reusable glass bottles at 15 hotels globally.
- Using energy from renewable resources at 30% of its hotels in India and 35% of its hotels in South India (IHCL, 2022).

(4) *ITC Hotels* in India have implemented various practices to reduce their carbon footprint and minimise environmental impact. These include using solar power, implementing energy-efficient lighting and HVAC systems and sourcing food from local farmers to reduce transportation emissions. ITC Hotels has introduced a recycling programme and has implemented sustainable waste management practices. ITC Hotels aims to reduce its carbon footprint and contribute to a more sustainable future by adopting these practices (GROUP, 2024).

(5) *RADISSON BLU hotels* in India have implemented various initiatives to reduce their carbon footprint for a sustainable future. These include using energy-efficient lighting and appliances, implementing recycling programs and sourcing food and supplies from local and sustainable suppliers. Radisson Blu hotels in India have introduced electric vehicle charging stations and promote the use of public transportation for guests (Hotel, 2023).

The research examines how technology improves the customer experience and encourages customers and stakeholders to use it. The study emphasises the advantages of implementing technology and sustainable practices. Recently, hotels have started incorporating digital practices to reduce their carbon footprint. Hotels can minimise paper usage using digital check-ins, room keys and billing methods. This lowers the need for paper, ink and printers, which contributes to reducing waste and carbon emissions. Another way is by utilising smart building technologies such as motion sensors and building automation systems that adjust lighting and temperature based on occupancy, thus reducing energy consumption. Hotels can also implement water conservation programs that minimise water usage, such as low-flow showerheads, toilets and laundry practices. Certain hotels are adopting renewable energy sources by utilising solar panels and wind turbines to power their premises. By implementing digital practices, hotels can reduce their carbon footprint, save on operational costs, and improve their reputation as environmentally responsible businesses.

Raising awareness, informing and sensitising internal and external stakeholders is crucial in reducing carbon footprint. The Following strategies can be adopted by companies: Developing a comprehensive communication plan can include various channels such as email, social media and internal newsletters. Creating engaging content to highlight the importance of reducing carbon footprint; hosting workshops, seminars and training sessions to educate stakeholders on sustainable practices and technologies and encouraging stakeholders to take the lead in reducing their carbon footprint by setting personal goals, tracking progress, celebrating successes and recognising individuals who have made significant contributions.

6. Conclusion

This research study examines the role of emerging technologies such as AI, big data analytics, IoT, and blockchain in addressing the carbon footprint. It has been observed that digitisation is a tool for achieving the objectives of reducing carbon footprints: it is also a central source of transformational change – the poor awareness level of stakeholders regarding carbon footprint results in the utilization of more resources. Academicians and researchers are creating awareness regarding responsible tourism management. The tourism industry needs to adopt responsible practices for optimising natural resources, which are manipulated to develop accommodation and recreation facilities. Therefore, tourism needs to be organised to overcome the obstacles of deforestation, overconsumption of land in hilly areas, melting glaciers and other concerns. There is a requirement for responsible business practices in the digital age.

There is a need to implement comprehensive guidelines to tackle the rising carbon footprint with strict compliance. Despite several international guidelines, a consistent and widely agreed-upon methodology for assessment must be needed. Climate change is an impending global emergency that poses significant environmental, socio-economic and geopolitical concerns. Rising temperatures, unpredictable rainfall, drought and cyclones are wreaking havoc on agricultural production, posing a severe threat to global tourism. Climate change creates more threats than opportunities by producing regional and seasonal fluctuations in visitor flows.

Hotels can reduce their carbon footprint in several ways.

- They can implement energy-efficient practices in their daily operations. This can include using LED lighting, installing motion sensors for lighting and HVAC systems and upgrading to energy-efficient appliances.
- Hotels can reduce waste by implementing recycling programs, using reusable items such as towels and linens and reducing single-use plastics.
- Hotels can source food and supplies locally, reducing transportation emissions and supporting local communities.
- Hotels can encourage and educate their guests on sustainable practices, such as turning off lights and electronics when not in use, using public transportation or walking instead of driving and participating in the hotel's sustainability initiatives.

By implementing these strategies, hotels can significantly reduce their carbon footprint and contribute to a more sustainable and environmentally friendly industry.

6.1 Practical Implications

The chapter recommends technology-supported experience concepts for reducing the carbon footprint at every step. The study will synthesise the most recent technological concepts to improve customer experience. It also provides insight into educating consumers about adopting sustainable practices.

6.2 Originality

The study attempted to address the benefits of green practices in providing a conducive environment nurtured with the appropriate practices to leave a lasting impression on tourists. It will also discuss how sustainable practices can spark a revolution.

7. Further Studies

Further research into carbon footprint reduction will assist in determining the industry's initiatives towards minimising environmentally harmful emissions and risks for long-term industrial development. A subsequent study could focus on the industry's emphasis on innovation in managing energy needs and climate impact. This research study emphasises the benefits and drawbacks of the digital revolution on sustainable practices.

References

Abeydeera, L. H. U. W., & Karunasena, G. (2019). Carbon emissions of hotels: The case of the Sri Lankan hotel industry. *Buildings*, *9*(11), 227.

Al, N., Aguiar, A. P., Boza-Kiss, B., Campagnolo, L., Chabay, I., Collste, D., Comolli, L., Gomez-Echeverri, L., Goujon, A., Grubler, A., Jung, R., Kamei, M., Kamiya, G., Kriegler, E., Kuhn, M., Leininger, J., Martin-Shields, C., Mayor-Rodriguez, B., Miller, J., & Rockström, J. (2019). *The digital revolution and sustainable development: Opportunities and challenges.*

Bănescu, C., Boboc, C., Ghiță, S., & Vasile, V., (2021, June 3–5). Tourism in the digital era. In: R. Pamfilie, V. Dinu, L. Tăchiciu, D. Pleşea, & C. Vasiliu (Eds.), *7th BASIQ international conference on new trends in sustainable business and consumption* (pp. 126–134). ASE. https://doi.org/10.24818/BASIQ/2021/07/016

Chan, E. S. (2021a). Influencing stakeholders to reduce carbon footprints: Hotel managers' perspective. *International Journal of Hospitality Management*, *94*, 102807.

Chan, E. S. W. (2021b). Why do hotels find reducing their carbon footprint difficult? *International Journal of Contemporary Hospitality Management*, *33*(5), 1646–1667. https://doi.org/10.1108/IJCHM-10-2020-1151

Cheng, X., Xue, T., Yang, B., & Ma, B. (2023). A digital transformation approach in hospitality and tourism research. *International Journal of Contemporary Hospitality Management*. https://doi.org/10.1108/IJCHM-06-2022-0679

Dodds, R., Bessada, T., Garcia, J., McDougall, A., & Thiesen, N. (2012). Carbon offsetting programs in North America: Assessing the involvement of the hospitality and tourism industry. *WIT Transactions on Ecology and the Environment*, *161*, 259–269. https://doi.org/10.2495/ST120211

Dubois, G., & Ceron, J. P. (2006). Tourism/leisure greenhouse gas emissions forecasts for 2050: Factors for change. *France Journal of Sustainable Tourism*.

GROUP, I. H. (2024, January 15). Responsible luxury. https://www.itchotels.com/in/en/responsible-luxury/sustainability

Hotel, R. (2023). Think Planet. https://www.radissonhotels.com/en-us/corporate/responsible-business/planet

Hu, A., Huang, C.-Y., Chen, C.-F, Kuo, C.H., & Hsu, C.-W. (2015). Assessing carbon footprint in the life cycle of accommodation services: The case of an international tourist hotel. *The International Journal of Sustainable Development and World Ecology*, *22*, 1–11. https://doi.org/10.1080/13504509.2015.1049674.

IHCL. (2022). https://www.ihcltata.com/press-room/ihcl-continues-endeavour-to-reduce-carbon-footprint-with-35-percent-of-its-hotels-in-south-india-utilizing-energy-from-renewable-source/

IHCL continues endeavour to reduce carbon footprint with 35 per cent of its hotels in south India utilizing energy from renewable sources. (2022). https://www.thelalit.com/wp-content/uploads/2021/01/We-Care-Booklet-Final.pdf

Jung, T. (2014). Augmented Reality (AR) in Urban Heritage Tourism. *e-Review of Tourism Research*, (5), 1–5.

Limited, E. (2024, February 13). Reducing our Carbon Footprint. https://www.eihltd.com/responsibility/environment/

Maitra, R. (2021). Adoption and implementation of digital transformation for the sustainability of tourism and hospitality business in India. *Journal of Services Research*, *21*, 89–111.

Maitra, R., Shukla, D., Janjua, L., & Raja, M. S. (2023). Implications of social media in sustaining renewable energy usage. In *Implications of industry 5.0 on environmental sustainability* (pp. 192–209). IGI Global.

Maitra, R., Tripathi, R., Kumar, A., & Shukla, D. (2023). Assessing the relationship between social media and sustainable. *Tourism Management*, *4*(1).

Miola, A. (2019). *The digital revolution and sustainable development: Opportunities and challenges*. https://doi.org/10.22022/TNT/05-2019.15913

Schubert, F., Kandampully, J., David, S., & Kralj, A. (2010). Exploring consumer perceptions of green restaurants in the US. *Tourism and Hospitality Research*. https://doi.org/10. 10.1057/thr.2010.17

Silva, R. G. (2003). *Introduction to augmented reality*. Technical report 25/2003. The Turkish city of Bursa: National Laboratory for Scientific Report.

Singh, N., Cranage, D., & Lee, S. (2014). Green strategies for hotels: Estimation of recycling benefits. *International Journal of Hospitality Management*, *43*, 13–22. https://doi.org/10.1016/j.ijhm.2014.07.006

Sun, Y. Y., Lin, P. C., & Higham, J. (2020). Managing tourism emissions through optimizing the tourism demand mix: Concept and analysis. *Tourism Management*, *81*, 104161.

Tiwari, S., Rosak-Szyrocka, J., & Żywiołek, J. (2022). Internet of Things as a sustainable energy management solution at tourism destinations in India. *Energies*, *15*(7), 2433.

Tong, Y., Zhang, R., & He, B. (2022, February 5). The carbon emission reduction effect of tourism economy and its formation mechanism: An empirical study of China's 92 tourism-dependent cities. *International Journal of Environmental Research and Public Health*, *19*(3), 1824. https://doi.org/10.3390/ijerph19031824

Truong, T. C. (2022). The impact of digital transformation on environmental sustainability. *Advances in Multimedia*. 6324325. https://doi.org/10.1155/2022/6324325

TWI2050 – The World in 2050. (2019). *The digital revolution and sustainable development: Opportunities and challenges. Report prepared by The World in 2050 initiative*. International Institute for Applied Systems Analysis (IIASA). www.twi2050.org

UNWTO. (2022, December 13). The travel and tourism sector commits to halting and reversing biodiversity loss by 2030. https://www.unwto.org/news/unwto-and-sustainable-hospitality-alliance-at-cop15

UNWTO. (2022, May 31). UNWTO world tourism barometer and statistical annex. https://www.e-unwto.org/doi/epdf/10.18111/wtobarometereng.2022.20.1.3?role=tab

Chapter 7

Sustainable Tourism Initiatives for Tribal Community Development in India

Vipin Singh[a,1] *and Jyoti Prakash*[b,2]

[a]Vellore Institute of Technology, India
[b]The Lalit Suri Hospitality School, India

Abstract

Sustainable tourism plays a crucial role in fostering community development, particularly in regions where indigenous tribal communities are economically disadvantaged despite their rich cultural heritage. India is home to diverse tribal populations with distinct traditions. This chapter explores the potential of sustainable tourism to drive community development in India's tribal regions. The key objectives include understanding the cultural significance of these communities as tourism assets, assessing the economic empowerment potential of sustainable tourism, evaluating environmental sustainability, exploring social inclusivity and scrutinising the role of policy and governance.

Keywords: Sustainability; green tourism; responsible tourism; tribal community; ecotourism

1. Introduction

Experience the transformative power of sustainable tourism! In today's world, we are all too familiar with the detrimental effects of mass- and overtourism. But fear not, for a brighter future is within reach. By embracing the concept of sustainable development, we can pave the way for a more harmonious and responsible approach to travel. Join us on this journey as we strive to create positive impacts

[1]https://orcid.org/0000-0001-5566-0151
[2]https://orcid.org/0000-0002-7892-6052

Sustainable Tourism, Part B, 123–134

Published under exclusive licence by Emerald Publishing Limited
doi:10.1108/978-1-83608-326-920241007

and preserve the beauty of our world. Together, we can make a difference! (Korstanje & George, 2020) Discover the undeniable truth: economies simply cannot flourish in a sustainable manner when the insatiable greed of capitalism is given free rein to amass boundless short-term profits, while conveniently shifting the burden of negative repercussions onto the unsuspecting public in the guise.

According to Koens et al. (2018), the effect of overtourism manifests itself as a consequence of a substantial surge in visitor numbers, leading to a state of overpopulation in locales where the indigenous populace is subjected to the adverse effects of transient and periodic tourism surges. According to Milano et al. (2019), the phenomenon in question exerts a profound and enduring influence on the daily routines, availability of resources and holistic welfare of the indigenous populace. This chapter delves into various facets of community development through tourism in tribal areas. Key objectives include understanding the cultural significance of these communities as tourism assets, assessing the economic empowerment potential of sustainable tourism, evaluating environmental sustainability, exploring social inclusivity and scrutinising the role of policy and governance.

As a result, it is more important than ever to work towards sustainable development employing cutting-edge industrial trends and to ensure a healthy environment and thriving service sector for future generations. Stakeholders, which include tour operators, travel brokers, hotels, guests and hosts, should be consulted throughout the development of green tourism. Green tourism is an initiative that promotes environmental preservation via the use of sustainable practices. The primary motivation for this is to ensure long-term stability in all aspects of society and the economy. Therefore, ecotourism is contingent upon their adherence to principles of environmental and ecological sustainability, as well as their economic viability. Ecotourism serves as a catalyst for engaging the local community in the noble pursuit of conserving the delicate balance of ecology and biodiversity within a given area. This symbiotic relationship between the preservation of biodiversity and the local community is further fortified by the provision of economic incentives, thereby fostering a harmonious coexistence. Ecotourism plays a pivotal role in the preservation of biodiversity, ensuring the continued prosperity of local communities. It necessitates conscientious conduct from both tourists and the tourism industry, while also fostering the growth of small and medium-sized enterprises within the sector. Moreover, it prioritises the minimal utilisation of natural resources and emphasises the active involvement, ownership and economic prospects for rural populations. Ultimately, ecotourism encompasses invaluable educational encounters, enriching the experiences of all involved parties (Iyer Kailash, 1995).

The exponential proliferation of mass tourism engenders deleterious consequences. The deleterious economic, social and environmental ramifications engendered by the development of tourism necessitate the imperative to embrace diverse strategies for sustainable tourism development. In this challenging situation, the sector used responsible tourism as a CSR approach. Destination operations depend on socio-economic and environmental conditions.

According to Markose et al. (2022), the surge in environmental consciousness initiatives underscores the imperative of transitioning towards a low-carbon epoch, thereby necessitating the contemplation of green tourism as the prospective trajectory for the advancement of the tourism industry. The cultivation of ecologically conscious tourism ought to be fostered through a comprehensive examination of the various stakeholders involved.

2. The Concern for Sustainable Tourism

Sustainable development means advancement that meets present needs without endangering the ability of future generations to meet their own needs (WTO & UNEP, 2005).

Throughout the annals of time, the veneration of nature and the adherence to principles of conservation have remained inextricably intertwined with the fabric of Indian philosophical and cultural ideologies. Traces can be traced back to the ancient civilisations of India, wherein individuals diligently cultivated the profound philosophy of the interconnectedness and unity of all living beings. The venerable Indian tradition has long espoused the notion that humanity is intricately intertwined with the natural world, thereby urging individuals to behold all manifestations of existence through the lens of profound affection and reverence.

It is lamentable that over the last several decades, there has been a convergence between the relentless pursuit of financial wealth and economic advancement in India and other regions and the exploitation of the natural environment in all its manifestations. Presently, the global community is confronted with a profound dilemma that poses a significant threat to its future prospects. The lush forest regions and diverse ecosystems have been consistently depleted in order to construct solid structures. The ongoing deforestation of forest reserves has resulted in the phenomenon of global warming and the subsequent greenhouse effects (Annamalai, 2007).

Fortunately, this has resulted in a significant realisation, prompting the global community to awaken to the need of assuming responsibility for the natural environment and paving the way for new opportunities and initiatives.

Sustainable development and tourism, as argued by Hall (2011) and Journeys (2023), present change, that is, the transition from one state or situation to another. Depending on how the change was first evaluated, the outcome of such a shift might be either beneficial to achieve extensive involvement and consensus building towards a goal of sustainable tourism, strong political leadership is essential. To do this, it is vital to keep an eye on the effects and take remedial action when it's called for. Keeping visitors happy is an important part of sustainable tourism, as is educating them on the importance of eco-friendly practices and encouraging them to adopt them throughout their travels (Thryambakam, 2013). Ecotourism as a means of achieving sustainable development is a major global problem. This idea has helped several nations guarantee growth in their regions. In the context of ecotourism development, 'dimensions' relate to the environmental, economic and social elements of tourist development. According to Bhuiyan, Md., et al. (2012), in order to ensure the long-term viability of sustainability development, a sufficient balance has to be formed between these dimensions.

3. Tribes in India

British colonialists in India invented the word 'tribe' to characterise an unwritten society based on blood ties that had yet to develop a recorded history. The majority of people who were not born into India's caste system were classified as members of a 'tribe' by the colonial government. The colonial authorities had designated several of these people groups as 'criminal tribes', much like the Dalits, who were either outcasts or shudras under the caste system, but this was corrected following Independence in 1949 (Singh,1982).

Experience the brilliance of the Indian Constitution as it unveils the true essence of Scheduled Tribes (ST). Delve into the depths of Articles 366(25) and 342, where the very definition of ST is meticulously crafted. Embrace the wisdom and foresight that shaped the destiny of a nation, as you explore the profound significance of these constitutional provisions. Introducing the extraordinary traits that define a community as ST: the captivating allure of 'primitive' qualities, the enchanting tapestry of a distinctive culture, the captivating seclusion of geographical isolation, the awe-inspiring trepidation of connecting with a vast population and the gripping sense of insecurity that permeates the air. Unveiling an unwritten truth, these time-honoured standards have gracefully stood the test of time, even though they may not have been explicitly inscribed within the sacred pages of the Constitution (Defending FoRB, n.d.). Unlock the treasure trove of historical wisdom! Immerse yourself in the profound insights of the 1931 Census, the groundbreaking First Backward Classes Commission reports from 1955 and the visionary Advisory Committee (Kalelkar) on the Revision of SC/ST lists from the esteemed Lokur Committee in 1965. But wait, there's more! Delve into the illuminating findings of the Joint Committee of Parliament on the Scheduled Castes and Scheduled Tribes Orders (Amendment) Bill from 1967. All these invaluable sources converge to shape the very essence of this definition. Introducing the captivating world of tribes, where the terms 'Tribe', 'Scheduled Tribe' and 'Adivasi' may seem similar at first glance, but hold unique and fascinating distinctions. Discover the captivating tribes of Northeast India, who possess a unique identity that sets them apart. Interestingly, these remarkable communities often choose not to identify themselves as 'Adivasi' (Government of India, 1931, 1955, 1965, 1967). Unlock the hidden potential of your group's ST status, carefully crafted and determined by the esteemed state governments. Discover the fascinating truth: individuals classified as STs within one state may not necessarily bear the same classification in another state (Bhagat, 2006).

4. Tribal Tourism: Tribes in India

Across 28 states and 2 territories, there are 705 unique tribal communities that have been recognised as STs. An estimated 10.43 crore (8.6% of the overall population) identify as tribal, and 89.97% of them live in rural regions, while just 10.03 crore (or 10.3% of the total population) call the cities home. When compared to the 17.69% decadal increase in the world's population between 2001 and 2010, the number of indigenous people rose by 23.66% between the 2001 and 2011 Censuses. Most ST's call the area around India divided into two distinct regions: the centre and the northeastern parts of the country (India-IWGIA, 2024).

The Central region of India harbours a significant proportion, exceeding 50%, of the nation's STs. These tribes are specifically concentrated in Madhya Pradesh, accounting for 14.69% of the total, followed by Chhattisgarh with 7.5%, Jharkhand with 8.29%, Andhra Pradesh with 5.7%, Maharashtra with 10.08%, Orissa with 9.2%, Gujarat with 8.55% and Rajasthan with 8.86% (Government of India, 1931, 1955, 1965, 1967). The Northeast constitutes a secondary geographical expanse encompassing Assam, Nagaland, Mizoram, Manipur, Meghalaya, Tripura, Sikkim and Arunachal Pradesh. Merely seven states within the vast expanse of India, namely Madhya Pradesh, Maharashtra, Orissa, Gujarat, Rajasthan, Jharkhand and Chhattisgarh, serve as the abode for a significant majority, surpassing two-thirds, of the nation's ST populace (Lusome & Bhagat, 2020).

5. Role of Sustainable Tourism in India

As human activities have an increasing effect on the environment, the relevance of sustainable tourism has grown in recent years. Some examples of how to put its tenet to work – that of reducing negative impacts while increasing favourable ones – are as follows:

- This helps to ensure that local nature, culture and environment are protected, while also allowing us to benefit from our use of natural and cultural resources without risking their destruction via inefficient usage. Sustainable travel is founded on the three R's: reduction, reuse and recycling.
- Sustainable tourism development enhances local communities by strengthening the local economy, leading to increased prosperity for everyone involved. To ensure that the economic benefits of tourism reach the local community, visitors need to support local businesses whenever possible.
- Preservation of cultural and ecological variety demonstrates respect for local populations and their practices. A sustainable tourism approach involves sharing cultures without forcing them and eliminating preconceptions about diverse cultures and faiths. Therefore, sustainable tourism preserves cultural and ecological assets for future generations.

6. Indian Tribal: Challenges and Issues

The majority of tribal communities exhibit underdevelopment across several domains, including social, political, economic, and educational aspects. There exists variation in the socio-economic and cultural landscape throughout different regions of India. The northeastern region has seen a prolonged period of heightened tension, whereas the mainland of central India faces several challenges including poverty, unemployment, indebtedness, underdevelopment and lack of education (Government of India, 1931, 1955, 1965, 1967). When compared to indigenous communities in other geographical areas, the tribes residing in the northeastern region exhibit a notable degree of political engagement, possess advanced literacy skills and maintain a commendable quality of life. The indigenous communities were forcibly displaced from their ancestral lands. The indigenous landholders saw a gradual displacement by the landowners and those involved in moneylending activities (Singh, 1982).

Burman (2003) posits that indigenous populations have a relative disadvantage in terms of literacy rates and economic development, rendering them comparatively less advanced. Tribal communities exhibit diverse characteristics across social, political, economic and ecological dimensions, resulting in different variations in the nature and extent of their respective difficulties. Visible divisions can be observed among different groups, such as hill tribes and plainsmen, settled farmers and those involved in forest-based economic activities and individuals who have converted to Christianity or Islam from Hinduism, as opposed to those who continue to adhere to traditional tribal practices. Despite these distinctions, individuals frequently face the following challenges: (1) The presence of exploitation and poverty, (2) the existence of economic and technological backwardness, (3) the experience of social and cultural disadvantages and (4) challenges related to integration within the non-tribal community.

As per the scholarly work of Dube (1988), it is posited that there exist five distinct classifications of indigenous communities, commonly referred to as tribal people. (1) Individuals who reside in seclusion; (2) individuals who engage with the neighbouring non-tribal populace while preserving their unique characteristics; (3) individuals who coexist within a village alongside caste groups, sects and religious establishments while upholding their distinct identity; (4) individuals who have been relegated to the marginalised position of untouchables and (5) individuals who enjoy elevated social, economic and political standing. One of the primary challenges encountered by indigenous communities in India pertains to the dearth of educational opportunities, resulting in limited access to academic institutions. Access to certain resources is restricted within the society due to a multitude of social taboos. These taboos encompass a wide range of factors, including deeply ingrained superstitious beliefs, prevailing poverty, a general lack of interest and limited availability of resources, among others. It is a prevailing occurrence within tribal communities for individuals to develop dependencies on alcohol and tobacco products, which are often regarded as integral components of their cultural customs and social rituals. In certain locales, individuals partake in the imbibing of a homemade libation known as Hadiya, which holds significance as a customary practice to alleviate hunger pangs. It has been purported that a substantial proportion, approximately 51%, of indigenous communities in India partake in the consumption of alcoholic beverages, regardless of whether they are produced domestically or procured from commercial establishments. This habitual practice has been observed to exert detrimental consequences on their overall well-being.

As per the scholarly work of Dube (1988), a comprehensive categorisation of tribal populations can be delineated into five distinct types. The first category encompasses those who reside in seclusion, detached from external influences. The second category comprises individuals who engage with neighbouring non-tribal communities yet diligently preserve their unique cultural attributes. The third category encompasses tribal groups that coexist within village settings alongside caste groups, sects and religious organisations, all the while upholding their distinct tribal identity. The fourth category pertains to those who have regrettably been relegated to the marginalised status of untouchables. Lastly, the fifth category encompasses tribal individuals who have attained elevated positions within society, enjoying favourable social, economic and political standing. One of the primary challenges encountered by indigenous communities in India pertains to the dearth of educational opportunities, resulting in limited accessibility to academic institutions. Access to certain resources is restricted within the society due to a multitude of social taboos. These taboos encompass a wide range of factors, including deeply ingrained superstitious beliefs, prevailing poverty, a general lack of interest and limited availability of necessary resources, among others. It is a prevailing phenomenon within tribal communities that the consumption of alcohol and tobacco products has become deeply ingrained, often regarded as a customary and social ritual. In certain regions, it is customary for individuals to partake in a locally crafted libation known as Hadiya, which serves as a ceremonial practice aimed at alleviating hunger pangs. It has been purported that a substantial proportion, approximately 51%, of indigenous communities in India partakes in the consumption of alcoholic beverages, be it in the form of homemade concoctions or commercially available products. Regrettably, this practice exerts a deleterious impact on their overall well-being and health. Furthermore, the exponential advancement of technology and the unparalleled dominance of global capitalism have facilitated the exploitation and depletion of natural resources within susceptible regions inhabited by indigenous communities. In areas where indigenous communities are subject to legal constraints, such as those pertaining to land ownership, forestry regulations, the utilisation of minor forest products and water resource management, their access to forest resources is effectively restricted. The availability of primary resources, such as fuel, fodder and minor forest products, which were previously accessible to the peasant class, has undergone a significant transformation. These resources have either become scarce or necessitate large-scale commercial imports in the present era. The phenomenon of globalisation has been intricately associated by various indigenous communities with the discernible escalation in financial expenditures, the erosion of occupational security and the dearth of attainability pertaining to healthcare services.

According to Khanday and Majeed (2021), the tribal populations residing in various areas have been the subject of multiple demonstrations as a result of forced relocation, which has occurred in the context of liberalisation, privatisation and globalisation (LPG). The forceful relocation of tribal groups in India has become an increasingly widespread and concerning phenomenon, resulting in the creation of significant infrastructure development projects. According to studies, a significant number of about 21 million persons have been internally displaced due to development endeavours, as seen among various tribal communities. Despite constituting a mere 8% of the overall population, indigenous communities account for almost 40% of the displacement resulting from infrastructure development. While there are some limitations in the process of rehabilitating indigenous villages affected by development projects such as the initiative '*Best Village Tourism Portal*' and Rural Tourism Portal launched by honourable Union Tourism Minister Shri. G. Krishna Reddy in the Global Tourism Investor Summit 2023 (Shri G. Krishan Reddy Launches 'Best Tourism Village Competition Portal', Global Tourism Investor Summit 2023 Portal and Rural Tourism Portal). Among the estimated 8.5 million individuals who have experienced displacement as a result of infrastructure development initiatives and natural calamities, a mere 2.1 million individuals belonging to indigenous groups have been provided with rehabilitation measures up to this point.

7. Recent Initiatives Taken in India

The Indian government's Tourism Ministry has recently taken a number of measures to boost the country's booming tourism sector. These include rolling out brand-new initiatives like Swadesh-Darshan and Prasad, revamping older programmes like hunar-se-rozgar tak, expanding the availability of electronic tourist visas to a wide range of countries and introducing innovative new services like mobile applications and a dedicated helpline for Incredible India's most ardent visitors (Ministry of Tourism, Government of India, 2024). Here are some further government programmes:

- Improving existing tourist facilities and building new ones from scratch: The government launched the Swadesh-Darshan strategy to promote tourism based on certain themes. Twenty-seven or more projects totalling Rs 2261cr have been granted for two States, all to develop world-class facilities to benefit cultural assets and to improve tourism attraction. The government's initiative to improve spiritual and pilgrimage locations is known as 'Prasad'.
- In February 2016, the Ministry launched a smartphone app called Swachh-Paryatan via which users may report unsanitary conditions at tourism attractions.
- The esteemed Ministry of Tourism, in its wisdom, graciously established a round-the-clock, cost-free Tourist Helpline in various languages during the auspicious month of February in the year 2016. This invaluable resource may be accessed by dialling the distinguished number 1800-11-1363. Not only does it encompass the Hindi and English languages, but it also incorporates an additional 10 international languages.
- The International Tourism Mart is an annual event that aims to display the untapped tourism potential of the Northeastern states. Its primary purpose is to promote and raise awareness about these regions' various attractions and opportunities.

- The Adarsh Smarak initiative, spearheaded by the Archaeological Survey of India (ASI), has successfully identified a multitude of historical monuments that possess immense potential for development, with the aim of transforming them into exemplary Model Monuments. The government shall endeavour to furnish the populace with essential amenities, such as Wi-Fi connectivity, robust security measures and inviolable spaces (Aneja, 2005). Furthermore, cinematic presentations shall be employed as a means to impart enlightenment regarding the significance of these historical landmarks, while simultaneously raising awareness about pivotal initiatives such as the Swachh Bharat Abhiyaan. A selection of the architectural marvels encompassed within this particular initiative includes the esteemed Leh Palace, the grandeur of Humayun's Tomb, the historic Red Fort, the enigmatic Elephanta Caves and the iconic Taj Mahal, among others.

Recently, our tourist minister Shri. G Kishan Reddy announced a plan to boost the tourist business in India. He also claimed the government would create six tourism circuits in the initial phase: Eco, Wildlife, Ramayan, Desert and Rural ones. Some of other major components under these plans are: (i) Building terminals for passengers, (ii) building of roads to connect tourist destinations, (iii) to use clean energy sources for lighting the streets, (iv) providing more emergency vehicles, repair and refuel facilities, (v) providing basic facilities such as water supply, sewerage, electricity and roads and (vi) upgrading communication systems like Wi-Fi, internet, mobile services (Press Information Bureau. (n.d.).

8. Sustainable Tourism in India: Major Issues and Challenges

Tourism may seem like a steam roller everywhere it goes. This may ruin natural areas. Ecotourism involves stopping and this quote contrasts ecotourism with mass tourism. Implementing an idea is significantly harder than creating it, hence it is realistic to expect 100% success in any implementation program. The same applies to ecotourism everywhere. India's ecotourism business is flourishing, yet it has challenges that are negatively impacting it. India's ecotourism concerns go against the notion.

- *Lack of awareness*
  Sustainable tourism is a term that is unfamiliar to many visitors all at once. To elaborate, he said, 'Many people who visit such facilities are sadly not totally aware of what "eco-tourism" truly is. What they need from a rental home isn't necessarily in line with what's meant by "eco-tourism" (Devesh, 2010)'.
- *Environmental issues*
  Sustainable tourism's promotion of the appropriation of uninhabited regions, such as national parks and tribal lands, is one of the industry's most worrisome aspects (Freya, 2018). Travellers that choose this route tend to come from more affluent socio-economic backgrounds, where they have the means to buy everything they choose. Travel companies give in to tourists' ever-increasing

expectations, and in order to accommodate and amuse them, they construct wholly artificial landscapes replete with mega-resorts, luxury hotels and shopping centres. Almost the whole ecosystem suffers as a result of these alterations; many plant and animal species lose their home, and others may become extinct as a result. When natural resources are used, ecosystems are altered and species go extinct. Changing the landscape for tourist purposes may have devastating effects on local wildlife populations. As resources are diverted from other areas of economic growth to support tourism, there is an 'opportunity cost' associated with this practice (Aguiñaga et al., 2018).

- *Socio-economic issues*
  Unfortunately, the economic gains are not always utilised for the conservation of the places or to help the local population since sustainable tourism businesses are frequently owned and controlled by outside interests, much like mass tourism. Furthermore, no attempt is made to estimate the capacity of host regions. Because of the overwhelming number of visitors during high season, public services often break down.
- *Impact over locals*
  Local communities gain from conservation, economic growth and environmental education via sustainable tourism. While parks and protected places immediately influence neighbouring residents, they also benefit from their protection. In two ways, sustainable tourism deceives communities. First, tourist monoculture replaces their economic and social framework. Furthermore, the commitment extended to them regarding the provision of a sustainable and continuous stream of revenue, as well as an equitable allocation of profits, is profoundly disingenuous. The indigenous population secures employment opportunities that are predominantly characterised by their inferior quality, diminished worth and temporary nature. They primarily function as purveyors of tourism guidance, sustenance or mementos within the immediate vicinity. The preponderance of revenue generated is predominantly allocated to the organisers' coffers. One could argue that the financial burden of ecotourism falls upon the local populace, while the primary beneficiaries of this enterprise are the prominent entities involved, such as the organisers, who reap substantial profits. The tax burden borne by tourists in India is notably the most substantial globally, in stark contrast to other Asian nations wherein the tax rates range modestly between 3% and 6%. Indian hotels impose a tax burden of approximately 40% upon their patrons. One must acknowledge that a significant determinant contributing to the constrained realisation of the nation's tourism potential lies in the regression of macroeconomic central planning, which has regrettably descended to the realm of micro-planning. Additional issues encompass substandard service, inadequate marketing efforts, deficient infrastructure, insufficient security measures, subpar governance, flawed execution of governmental directives, inadequate synergy between central and regional authorities in fostering tourism, excessive reliance on intermediaries and a conspicuous disparity between policy formulation and practical implementation, among others (Bhattacharya, 2023).

References

Aguiñaga, E., Henriques, I., Scheel, C., & Scheel, A. (2018). Building resilience: A self-sustainable community approach to the triple bottom line. *Journal of Cleaner Production*, *173*, 186–196.

Aneja, P. (2005). Tourism growth in India. *Kurukshetra*, *17*(9), 11–14.

Annamalai, M. (2007). Challenges and Changes in Indian Tourism. *South Asian Journal of Socio-Political Studies (SAJOSPS)*, *37*(1), 103–106.

Bhagat, R. B. (2006). Census and caste enumeration: British legacy and contemporary practice in India. *Genus*, *XII*(2), 119–134.

Bhattacharya, A. (2023). Sustainable tourism: Lack of awareness and other challenges. *The CSR Journal*. https://thecsrjournal.in/sustainable-tourism-awareness-challenges/. Accessed on October 9, 2023.

Bhuiyan, Md., A. H., Chamhuri, S., Ismail, S. M., & Islam, R. (2012). The role of ecotourism for sustainable development in east coast economic region (ECER), Malaysia. *OIDA International Journal of Sustainable Development*, *3*(9), 53–60.

Burman, B. K. R. (2003). Indigenous and tribal peoples in world system perspective. *Studies of Tribes and Tribals*, *1*(1), 7–27.

Chin, C. H., Chin, C. L., & Wong, W. P. M. (2018). The implementation of green marketing tools in rural tourism: The readiness of tourists? *Journal of Hospitality Marketing & Management*, *27*, 261–280.

Defending FoRB. (n.d.). Religious freedom and tribal rights in India. https://defendingforb.org/en/india/religious-freedom-and-tribal-rights-in-india/. Accessed on July 15, 2024.

Devesh, C. (2010). Tourism in India: Ensuring buoyancy and sustainability. *Yojana*, *13*(8), 16–18.

Dube, S. C. (1982). The study of complex cultures: Concepts and methodological issues. In *Understanding change: Anthropological and sociological perspectives*. Vikas Publishing House.

Dube, S. C. (1988). *Modernization and development: The search for alternative paradigms*.

Freya, H. (2018). Sustainable tourism: Sustaining tourism or something more? *Tourism Management Perspectives*, *25*, 157–160.

Government of India. (1931). Census of India. https://censusindia.gov.in/nada/index.php/catalog.31806. Accessed on July 15, 2024.

Government of India. (1955). First backward classes commission report. https://socialjustice.gov.in/writereaddata/UploadFile/78251687858425.pdf. Accessed on July 15, 2024.

Government of India. (1965). Advisory committee on the revision of SC/ST lists (Kalelkar Committee). https://socialjustice.gov.in/public/ckeditor/upload/11301676262859.pdf. Accessed on July 15, 2024.

Government of India. (1967). Joint committee of parliament on the scheduled castes and scheduled tribes orders (Amendment) bill. https://main.sci.gov.in/jonew/judis/4678.pdf. Accessed on July 15, 2024.

Hall, C. M. (2011). Policy learning and policy failure in sustainable tourism governance: From first- and second-order to third-order change? *Journal of Sustainable Tourism*, *19*(4–5), 649–671.

India - IWGIA (International Work Group for Indigenous Affairs). (2024). International Work Group for Indigenous Affairs. https://www.iwgia.org/en/india.html. Accessed on June 20, 2024.

Iyer Kailash, H. (1995). *Tourism development in India* (pp. 90–113). Vista International Publishing House.

Journeys, T. (2023). A guide to sustainable tourism in India. *Traverse Journeys – Travel That Transforms*. https://traversejourneys.com/blog/a-guide-to-sustainable-tourism-in-india. Accessed on October 9, 2023.

Khanday, S. A., & Majeed, I. (2021). Challenges encountered by tribal population in the era of globalization. *Research Journal of Humanities and Social Sciences*, *10*(2), 2321–5828.

Koens, K., Postma, A., & Papp, B. (2018). Is overtourism overused? Understanding the impact of tourism in a city context. *Sustainability*, *10*(12), 1–15.

Korstanje, M., & George, B. (2020). Demarketing overtourism: The role of educational interventions. In A. C. Yallop & H. Séraphin (Eds.), *Overtourism and tourism education* (pp. 81–95). Routledge.

Lusome, R., & Bhagat, R. B. (2020). Migration in Northeast India: Inflows, outflows and reverse flows during pandemic. *Indian Journal of Labour Economics*, *24*, 1–17.

Markose, N., Tazhathethil, B. V., & George, B. (2022). Sustainability initiatives for green tourism development: The case of Wayanad, India. *Journal of Risk and Financial Management*, *15*(2), 1–13.

Milano, C., Novelli, M., & Cheer, J. M. (2019). Overtourism and tourismphobia: A journey through four decades of tourism development, planning and local concerns. *Tourism Planning & Development*, *16*(4), 353–357.

Ministry of Tourism, Government of India. (2024). https://tourism.gov.in/sites/default/files/2024-02/usq.292%20for%2005.02.2024.pdf. Accessed on July 15, 2024.

Press Information Bureau. (n.d.). https://pib.gov.in/PressReleasePage.aspx?PRID51899949#:;:text5Kishan%20Reddy%20launched%20important%20initiatives,also%20launched%20on%20the%20occasion%20. Accessed on October 9, 2023.

Shri G. Kishan Reddy launches 'Best Tourism Village Competition Portal', Global Tourism Investor Summit 2023 Portal and Rural Tourism Portal in New Delhi today. (n.d.). https://pib.gov.in/PressReleasePage.aspx?PRID=1899949#:~:text=Kishan%20Reddy%20launched%20important%20initiatives,also%20launched%20on%20the%20occasion. Accessed on 24 February 2023.

Singh, K. S. (1982). Transformation of tribal society: Integration vs assimilation. *Economic and Political Weekly*, *17*(34), 1376–1384.

Thryambakam, P. (2013). Impact of tourism on community development and sustainability in Maredumilli, East Godavari District of Andhra Pradesh – A case study. *Global Journal of Management and Business Studies*, *3*(9), 1013–1020.

WTO & UNEP. (2005). *Making tourism more sustainable: A guide for policy makers*. World Tourism Organization & United Nations Environment Programme. https://www.unep.org/resources/report/making-tourism-more-sustainable-guide-policy-makers

http://tourism.gov.in/sites/default/files/2020-01/Document.pdf. Accessed on 9 October 2023.

https://sustainabledevelopment.un.org/content/documents/12023Kulshreshtha.pdf. Accessed on 9 October 2023.

Chapter 8

Vibrant Village Programme Performance Measurement: The Relationships Among Event Performance, Tourist Revisit Intention and Word of Mouth

Rajinder Kumar[a,1], *Hamid Abdullah*[b] *and Tsering Chusket*[a]

[a]University of Ladakh, India
[b]Atal Bihari Vajpayee Vishwavidyalaya, India

Abstract

The Vibrant Village Programme (VVP) is a paramount step by the Government of India to develop sustainable communities through border tourism along the Indo-China border. The programme was launched with manifold objectives, e.g. border security (BS), event promotion (EP), infrastructure development (ID), livelihood creation (LC), check on out migration (OM) and local resource conservation (LRC). The programme performance was measured on the bases of six parameters, i.e. EP, LC, OM, LRC, service quality, BS and ID. The data was collected through a survey from participants of the *Last Run*, on 20th February 2023 at Pangong Tso (4225 Mt), Ladakh, India. The 65-survey data were obtained via a structured questionnaire from domestic tourists (who participated in the Last Run) on 9 parameters. The variance-based structural equation modelling (VB-SEM) was opted to test the hypothetical paths. This research was conducted with four objectives, i.e. (1) to put forward a programme performance measurement framework, (2) to find the mediation effect of word of mouth, among the relationship of Vibrant Village Programme Performance (VVPP) and destination revisit intention, (3) important Performance Metrix Analysis of the Model and (4) socio-economic impacts of the event in the border area. Tourism stakeholders can find important knowledge and can facilitate

[1]https:orcid.org/0000-0003-0466-1140

Sustainable Tourism, Part B, 135–163

Published under exclusive licence by Emerald Publishing Limited
doi:10.1108/978-1-83608-326-920241008

holistic destination management (especially event performance). The Smart PLS 4 was used in data analysis. The performance measurement of this programme is useable for policymakers, destination planners, researchers and business houses. This research work is a ready-to-implement document type; for tourism policymakers and managers, destination planners and border tourism event managers.

Keywords: Border tourism; sustainable communities; quality of services; word of mouth; Ladakh; event performance; structural equation modelling

1. Introduction

Events are an important motivator of tourism and have important role in destination development and its marketing, which is helpful in destination competitiveness. Events have a vital role for a destination as attractions, catalysts, animators, marketing, new tourism products, all-season destination and image builders (Getz et al., 1998; Getz, 2008; Janiskee, 1996; Ryan et al., 1998). But, prior to 1979, use of events as tourism attraction was not in practice (Gunn, 1979). Ritchie (1984, p. 2) defined events as 'major one-time or recurring events of limited duration, developed primarily to enhance awareness, appeal and profitability of a tourism destination'. Events can be online (virtual) and offline, small and large scale, indoor and outdoor (Aguilera et al., 2016; Kusonwattana et al., 2022). Events have diverse category including business to business events, e.g. World Travel Market London (Fragkogianni, 2018).

Events have an important role in community-building, urban renewal, cultural development, fostering national identities, regional economic and sociocultural development (Getz, 2005, 2008; Whitford, 2004). The suppliers create events with the purpose of developing and marketing destinations, promoting events, attracting tourists during lean seasons, infrastructure development, fostering a positive destination image, quality of services (QoS) and life (Connell et al., 2015; Getz, 2008). Getz (2005) explored seven typologies of planned events, i.e. cultural celebrations, political and state, arts and entertainment, business, and trade, educational and scientific, sports competition, recreational and private events.

1.1 Vibrant Village Programme (VVP)

The Central government has allocated 4800 crore for Vibrant Village Programme (VVP) from 2022 to 2026 year (Ministry of Home Affairs, 2023b), and events are created for a purpose to achieve strategic goals with an economic, environmental and cultural outcomes (Getz et al., 2007; Getz, 2008); last run was a planned event under VVP with the purpose of strengthening border security (BS), programme promotion, infrastructure development in border areas, livelihood creation in border villages, local resource conservation and stop outmigration from border areas to city (Ministry of Home Affairs, Government of India, 2023c). The VVP is accomplishing many strategic goals, and events are created to fulfil many strategic goals (Getz, 2008). Getz (2005) explored sports competition as event.

1.2 Frozen Pangong Lake Marathon in Ladakh

Ladakh's 21 km marathon, started from Lukung and ended at Maan village. Seventy five selected athletes from India and abroad have participated in the run, giving them an opportunity to make a bid for a Guinness world record for the world's highest frozen-lake marathon. India's first 'frozen-lake marathon' at Ladakh's Pangong Tso at a height of 13,862 feet was held on 20 February 2023. Ladakh hosted India's first frozen-lake marathon 'Last Run' at Pangong Tso. The marathon is being called the 'Last Run' to highlight the issue of climate change. According to Getz event typology, Last Run falls under Sports Competition category and the performance of this event was measured through the perspective of the demand side. Last Run was a sports event under VVP with multifaceted objectives (Ministry of Home Affairs, Government of India, 2023c). The Last Run was a Marathon on Frozen Lake, and it was named so because maybe next year that lake won't freeze because of climate change this event was a Guinness Book record (Administration of Union Territory of Ladakh, 2023). Last run was an example of good cooperation among event organiser, Leh administration, Indian Army, Indian Tibet Border Police and local community. The Walo et al. (1996) has explored that economic benefits are an output of good cooperation among event stakeholders.

1.3 Attributes of Vibrant Village Programme

1.3.1 Border Security and Tourism (BS)

India shares 15,106.7 Km of land border with neighbouring countries (Ministry of Home Affairs, 2023a). The border areas are considered economically marginalised and geographically isolated (Wastl-Walter et al., 2003). The promotion of border tourism can stimulate the economic activities in border areas (ETTravelworld, 2022). del Río et al. (2017) defined border tourism as 'temporary displacement of people to the dividing line between two countries contiguous areas'. Security is an important concern in border destinations (Bringas & Verduzco, 2008) and it has three dimensions population, regional and national security. Also, successful organisations of events enhance the trio's viability and strengthen national sovereignty (Gelbman & Timothy, 2011).

1.3.2 Event Promotion (EP)

The event has significant reach on different social media (Administration of Union Territory of Ladakh, 2023). Özdemir and Çulha (2009) assessed the event performance on through seven parameters and information reach among visitors through signboard were one of them.

1.3.3 Infrastructure Development (ID)

The VVP has multifaceted objectives including infrastructure development at border areas. This infrastructure includes roads, healthcare, Education opportunities, agriculture productivity, sports events, drinking water and sanitation, community upliftment and establishment of small-scale industries (Ministry of Home Affairs, 2023c).

1.3.4 Livelihood Creation (LC)

The focal point of VVP is livelihood creation along the border area through tourism promotion and sports events (Ministry of Home Affairs, 2023a).

1.3.5 Migration and Ladakh (OM)

Border migration can be formal and informal, and the reasons are healthcare, leisure, trade or work (Hussain, 2022; Namgail & Chaudhary, 2023; Tapia Ladino, 2023). Migration can be cross border and internal (Kone et al., 2018; Tapia Ladino, 2023). The migration in the Ladakh region is not cross border; because it's a close border and it restricts migration (Vezzoli, 2021) but internal. Migration in Ladakh is specifically a rural to urban form of migration of nomadic pastoralists (Goodall, 2004; Namgail & Chaudhary, 2023). The reasons for migration are quality of life, education, employment, healthcare climate change, border conflicts, concentration of tourism industry in Leh, shrinking pastoral and receding glaciers (Hussain, 2022; Namgail & Chaudhary, 2023; Williams-Oerberg, 2016). The VVP was launched by government of India to generate employment at border areas to address the migration (rural to urban) of locals through tourism (Ministry of Home Affairs, 2023b).

1.3.6 Local Resource Conservation (LRC)

Cultural immersion is an important experience for event participants through different activities at event destination (Mendes et al., 2013). Tourism is an important source of income in Ladakh (Namgail & Chaudhary, 2023). Ladakh has diverse ethnic groups and the land of multicultural history (Rowold et al., 2016). The VVP has an important motive to conserve the local culture, for creation of livelihoods opportunities in selected villages in border areas of Ladakh (Ministry of Home Affairs, 2023b). Promotion of tourism is an important tool to achieve the objectives of VVP.

1.4 Event Performance or Assessment Measurement

Performance includes the quantification of output of an event (Neely et al., 2002). Performance is an important indicator to measure the output level of an event and it has critical role in the success of an event as well (Chen et al., 2014; Crompton & Love, 1995). Bandi Tanner et al. (2018) explored seven fundamental indicator of event performance, i.e. size of an event, economic output, touristic importance, product innovation, networking among event stakeholders, sociocultural exchange and environmental impact. Also, Chen et al. (2014) measured the event performance on three dimensions (hedonic, event design and informative) of event. Özdemir and Çulha (2009) assessed the event performance on through seven parameters, i.e. event content (activities), staff (e.g. respond quickly), facilities (e.g. adequate area), food (e.g. traditional food), souvenirs (quality), convenience (restroom) and information (signboard). Funchal and Pinto (2018) explored the relationship between corporate event performance and governance. Globally, events at a destination are assessed or studied in context of stakeholder knowledge networks (Stokes, 2004), policy and programme (Whitford, 2004), portfolio (Getz, 2005), tourism plan and event policy (Getz et al., 2007) but most are based on marketing or economic impact (Getz, 2008). Also, events can be assessed in the context of demand and supply. A demand perspective can have assessment parameters like who travels for events and why; what they do and spend. The researcher called event participants as *event tourists* (Getz, 2008). Event tourists travel to event for work, leisure and cultural experience (Getz et al., 2007). Bohlin studied events and explored *travel decline with distance* (2000) and Prabha et al. (2006) studied this spatial-temporal phenomena through two parameters, i.e. travel cost to an event destination and it was used to measure an event's economic value. This study opted to explore the event performance of Last Run and it was organised under VVP. Hence, VVP was measured under objective success and subjective success of the event. There is a dearth of scholarship on event performance measurement at global platform and no research was found in Indian context with key word research '*event performance*' on SCOPUS database till September 2023.

2. Events Performance (EP) and Word of Mouth (WoM)

Word of Mouth (WoM) is most influential marketing activity among all the marketing initiatives (Fragkogianni, 2018). The WoM is a positive output about services consumed from a satisfied customer (Özdemir et al., 2016). Event performance is a specific motivation towards tourists' behaviour, e.g. WoM (Ajzen, 1991). The researcher explored that event performance has impact on WoM (Prado-Gascó et al., 2017). WoM is a subjective parameter of event performance. Özdemir and Çulha (2009) explored the connection among EP and WoM. WoM was measured through two items adapted sharing experience and recommendation of destination (Zeithaml et al., 1996).

2.1 Events Performance (EP) and Quality of Services (QoS)

According to Prado-Gascó et al. (2017), in sports event service experience is highly important and mentioned that event performance is linked to revisit intention (repurchase). QoS is a subjective parameter of event performance.

2.2 Events Performance (EP) and Revisit Intention (RI)

The repeat visit to a destination willingly by an individual is defined as RI (Tosun et al., 2015). The survival of tourism sector is dependent on RI of tourists (Ngoc & Trinh, 2015) and lesser the marketing cost (Kim et al., 2013; Loi et al., 2017). Events are catalyst in tourists revisit (McHone & Rungeling, 2000; Quinn, 2006). The RI of an event has a strong relationship with participant loyalty towards the event (Cronin & Taylor, 1992; Zeithaml et al., 1996). In addition, the event satisfaction also leads to revisit of the participants (Oliver, 1999; Yoon et al., 2010). According to Ajzen (1991), event performance is an important motivation towards participants' behaviour, e.g. revisiting a destination. According to Yuan et al. (2008) event performance has a dominant role in revisit of participants. Prado-Gascó et al. (2017) explored that the performance of event can be analysed objectively (out of the event) and subjectively (QoS, RI). Also, studies are focused on subjective performance variables. Johnston and Clark (2008), Özdemir and Çulha (2009) and Fragkogianni (2018) also explored the relationship between RI and EP. Besides, QoS and WoM have an effective relation (Prado-Gascó et al., 2017). RI includes positive words, keep attending and recommending an event (Cronin & Taylor, 1992; Zeithaml et al., 1996) (Fig. 8.1).

2.3 Events and Demographics

Event tourism market is complex as different event have different target market and different goals (Rachael & Douglas, 2001). Male and female event attendees

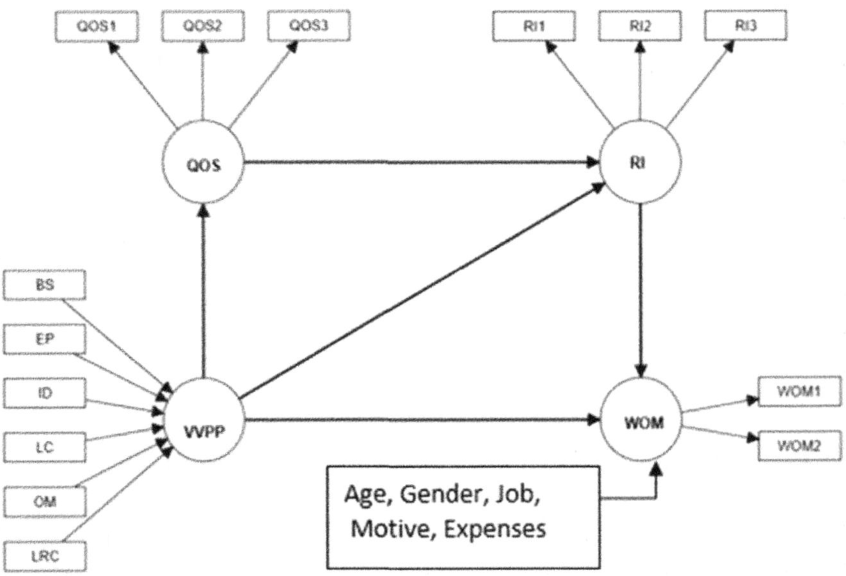

Fig. 8.1. Conceptual Framework. *Source:* Researcher.

have different socio-demographics, travel behaviour and motives (Kruger & Saayman, 2015). Size of the event was measured through length of stay of all event participants; economic value was measured through spending by each participant (Bandi Tanner et al., 2018).

3. Sample Size and Data Collection

In the present study, a total of 65 pan Indian domestic tourists were requested to participate in the survey. The vibrant village programme performance (VVPP), QoS, RI and WoM were the major parameters of questionnaire. The respondents answered the questions themed on VVPP, QoS, RI and WoM. The research study was descriptive in nature and validated on 65 domestic tourists travelling to Ladakh to participate in Last Run; India's first frozen-lake high-altitude marathon held on 20th February 2023. Thus, a total of 67 tourists were approached, and we have received 65 responses, representing 97.01% response rate. There were 75 participants in the marathon from across the world (Union Territory Administration of Union Territory of Ladakh, 2023). Thus, 65 questionnaires were used to draw interpretations in final analysis. The '10-times rule' method (Hair et al., 2011) incorporated in this study, and the minimum sample size as per this method was 30. Therefore, sample of 65 was found to be appropriate. Besides, this a telephonic interview was conducted after 9 months of event among organisers and villagers (6) to explore the socio-economic impacts of this event in border villages. It was more than 10% of 95 (local). The questions were based on registration fee, no. of participants, no. of homestay operational, no. of local participating in event management and business days during edition 1 and edition 2 of Last Run.

3.1 Scale Development

The VVPP was measured by BS, EP, ID, LC, OM and LRC. The parameters are adopted from VVP initiated by Government of India. All items were measured on a 5-point Likert scale (1 = strongly disagree, 5 = strongly agree). Details of the items are listed in Table 8.1.

3.2 Statistical Methods

Using the variance-based partial least square approach (PLS-SEM) in the Smart PLS 4 software, non-parametric structural modelling was utilised to assess and test the hypotheses (Ringle et al., 2015). In the study, PLS-SEM was determined to be an effective multivariate data analysis technique because it uses a complicated model (Hair et al., 2020). Also, the data analysis in this study includes a prediction perspective of WoM based on formative predictor variables of VVPP (Cheah et al., 2019; Sarstedt et al., 2019) in context of Ladakh which hosted Last Run and the event was recorded in Guinness Book record of the world. For prediction orientation in a theoretical framework for social and behavioural

Table 8.1. Construct, Item and Description.

Constructs	Item	Description
Border Security (BS)	BS1	In my opinion, Last Run is a best example of civil and defense cooperation.
	BS2	In my opinion, Last Run is a strategic move towards border dominance.
	BS3	In my opinion, last run will help in strengthening of border security.
Event Promotion (EP)	EP1	In my opinion, the Last Run is well promoted on government websites and newspapers/TV/radio.
	EP2	In my opinion, 'Last run' is well promoted on social media.
	EP3	In my opinion, there is sufficient availability of information at destination.
Infrastructure Development (ID)	ID1	How do you rate the acclimatisation facilities/Medical facilities in Ladakh.
	ID2	In my opinion, such events near border will strengthen accommodation infrastructure.
	ID3	In my opinion, such events near border will strengthen road infrastructure.
Livelihood Creation (LC)	LC1	This event will create job opportunities for unskilled villagers too.
	LC2	In your opinion, 'Last Run' will generate new tourism jobs in Ladakh.
	LC3	How do you rate the importance of Last Run towards the local economy?
Out Migration (OM)	OM1	How do you rate the importance of Last Run, to stop migration of Local people during winters.
	OM2	How do you rate the importance of Last Run to stop migration of local people to urban places.
Quality of Services (QoS)	QoS1	How do you rate the quality of transport to the event Pangong?
	QoS2	How do you rate the quality of catering facilities at Pangong?
	QoS3	How do you rate the quality of accommodation at Pangong?

Table 8.1. *(Continued)*

Constructs	Item	Description
Revisit Intention (RI)	RI1	I want to revisit Ladakh with my family.
	RI2	I want to visit Ladakh during all season.
	RI3	In my opinion Ladakh is worth revisiting Place
Local Resource Conservation (LRC)	LRC1	Last Run has used local resources in this event to ensure long-term success, also for the generations to come
	LRC2	Last Run, will help to promote the culture of Ladakh
	LRC3	Last Run, will help to preserve the cultural heritage in Ladakh.
Word of Mouth (WoM)	WoM1	I will recommend Ladakh to my friends/ family.
	WoM2	I will share only positive experiences about Ladakh with my family/friends.
Global Item	GI	Last Run was a value for money event.

Source: Researcher.

sciences, PLS-SEM is a suitable method (Hair et al., 2019). PLS-SEM was also appropriate in this investigation because operationalising advanced models and analysing mediation effects were additional study goals (Saari et al., 2021). To move on with 10,000 bootstrap subsampling with bias adjusted percentile technique (two-tailed test) for structural model assessments, the first-order and second-order measurement model assessments were evaluated first (Hair et al., 2022). The study is based on reflective-formative type higher order construct and repeated indicators use is there. Two methods can be incorporated to analyse the repeated indicators approach (Wetzels et al., 2009), embedded two-stage approach (Ringle et al., 2012) and the disjoint two-stage approach (Agarwal & Karahanna, 2000; Becker et al., 2012). The latter approach was opted in this research. As both approaches lead to similar results (Cheah et al., 2019) and there was no compelling reason to choose one over other.

4. Results and Analysis

4.1 Descriptive Statistics

4.1.1 Respondents' Profile

It was interested to explore the demographic profile of respondents. Table 8.2 shows that approximately 96.9% of respondents were male and only 3.1% were female,

Table 8.2. Demographic Profile of Domestic Tourists.

Variable		n	%
Gender	Male	63	96.9
	Female	2	3.1
Personality	Pro-environment	26	40.0
	Situational	4	6.2
	Adventure	35	53.8
Age	18–24	2	3.1
	25–34	22	33.8
	35–44	19	29.2
	45–54	20	30.8
	55–64	2	3.1
Travel motivation	Last Run Participant (LRP)	22	33.8
	Spectator	2	3.1
	Leisure	6	9.2
	Business	7	10.8
	Accompanying LRP	5	7.7
	LR is Guinness book record event	4	6.2
	Climate change message event	19	29.2
Job/profession/social status	Self-employed	6	9.2
	Business person	14	21.5
	Government employee	14	21.5
	Sports person	10	15.4
	Private job	8	12.3
	Unemployed	2	3.1
	Other	11	16.9
Visited Ladakh before	Yes	40	61.5
	No	25	38.5
How many times you have been to Pangong?	First time	22	33.8
	Second time	19	29.2
	Third time	12	18.5
	More than three times	12	18.5
How did you come to know about Last Run?	Friends	19	29.2
	Government website	6	9.2

Table 8.2. *(Continued)*

Variable		**n**	**%**
	Insta Handle of Travel Agency	14	21.5
	Other	26	40.0
Days	7<=10	36	55.38
	>10 < 16	29	44.61
Spending per day (₹)	>=2000 <=5000	49	75.38
	6000	16	24.61

Source: Researcher.

having varied social status, e.g. business person and government employees (21.5% each), sports person (15.4%), private job (12.3%), self-employed (9.2%) and unemployed (3.1%). It was found that 53.8% tourists tagged themselves as adventure loving personality while 40% happily accepted them as pro-environment tourists. In context of age parameter, majority of the tourists, i.e. 33.8% were in the age group of 25–34 and least number was from 18–24 and 55–64 age group, i.e. 3.1% only. It was explored that largest tourist, i.e. 33.8% travelled to Pangong, Ladakh with a travel motive to participate in Last Run (a high-altitude marathon), followed by 29.2% to witness an event which was pro-environment and only 3.1% travelled to spectate the event. The study revealed that 61.5% tourists revisited Ladakh and 38.5% were first time in Ladakh. It was found that 33.8% tourists were first time in Pangong. The government website reached to convey the message of Last Run among 9.2% only, 29.2% came to know about Last run from friends and 21.5% came to know through Instagram handle of a travel agency. The majority (55.38) of the tourist spent at least 7 days and maximum 10 days (7<=10) during their tour to Ladakh. A majority of 75.38% of the tourist spent at least 2,000 per day and maximum of 5,000 per day.

4.1.2 Criteria and Item Ranking

Table 8.3 shows the mean scores for the 17 VVPP criteria items, 4 QoS performance criteria items, 2 WoM and 3 RI performance criteria items. The range of mean ranking scores was from 4.523 (In my opinion, Last Run is a best example of Civil and defence cooperation, BS1) to 2.734 (I want to revisit Ladakh with my family, RI1). The criteria with the highest mean rankings were Border Security (M = 4.338), Infrastructure Development (M = 4.334), Out Migration (M = 4.285), Livelihood Creation (M = 4.210), Word of Mouth (WoM) (M = 4.123), Event Promotion (M = 4.0), Revisit Intention (M = 3.892), Local Resource Conservation (M = 3.867) and lowest was Quality of Services (M = 3.841).

Table 8.3 Construct, Item, Description, Mean.

Constructs	Item	Description	Mean
Border Security (BS) (*M* = 4.338)	BS1	In my opinion, Last Run is a best example of civil and defence cooperation.	4.523
	BS2	In my opinion, Last Run is a strategic move towards border dominance	4.292
	BS3	In my opinion, last run will help in strengthening of border security.	4.2
Event Promotion (EP) (*M* = 4.)	EP1	In my opinion, the Last Run is well promoted on government websites and newspapers/TV/radio.	3.8
	EP2	In my opinion, 'last run' is well promoted on social media.	4.138
	EP3	In my opinion, there is sufficient availability of information at destination	4.062
Infrastructure Development (ID) (*M* = 4.334)	ID1	How do you rate the acclimatisation facilities/Medical facilities in Ladakh	4.262
	ID2	In my opinion, such events near border will strengthen accommodation infrastructure.	4.277
	ID3	In my opinion, such events near border will strengthen road infrastructure.	4.462
Livelihood Creation (LC) (*M* = 4.210)	LC1	This event will create job opportunities for unskilled villagers too.	4.138
	LC2	In your opinion, 'Last Run' will generate new tourism jobs in Ladakh.	4.169
	LC3	How do you rate the importance of Last Run towards the local economy?	4.323
Out Migration (OM) (*M* = 4.285)	OM1	How do you rate the importance of Last Run, to stop migration of local people during winters.	4.292
	OM2	How do you rate the importance of Last Run to stop migration of local people to urban places.	4.277
	QoS1	How do you rate the quality of transport to the event Pangong?	3.892

Table 8.3 *(Continued)*

Constructs	Item	Description	Mean
Quality of Services (QoS) (M = 3.841)	QoS2	How do you rate the quality of catering facilities at Pangong	3.631
	QoS3	How do you rate the quality of accommodation at Pangong?	4
Revisit Intention (RI) (M = 3.892)	RI1	I want to revisit Ladakh with my family.	3.585
	RI2	I want to visit Ladakh during all season.	3.969
	RI3	In my opinion Ladakh is worth revisiting Place	4.123
Local Resource Conservation (LRC) (M = 3.867)	LRC1	Last Run has used local resources in this event to ensure long-term success, also for the generations to come	3.969
	LRC2	Last Run will help to promote the culture of Ladakh	3.646
	LRC3	Last Run will help to preserve the cultural heritage in Ladakh.	3.985
Word of Mouth (WoM) (M = 4.123)	WoM1	I will recommend Ladakh to my friends/family.	4.369
	WoM2	I will share only positive experiences about Ladakh with my family/friends.	3.877
Global item (M = 4.2)	GI	Last Run was a value for money event.	4.2

Source: Researcher.

4.2 Data Analysis and Results

Measurement (outer model or lower order model) and structural (inner model or SEM) model assessment are two essential parts of structural equation modelling. This study has reflective-formative model; lower order construct (reflective construct) and higher order construct (formative construct) are integral part of the study. The lower order construct assessment was carried through; factor loading, construct reliability (average variance extracted or explained, i.e. AVE); internal consistency of responses (composite reliability); discriminant validity of items (cross loading); and discriminant validity of construct through heterotrait-monotrait (HTMT) values. In reflective-formative model, assessment of convergent validity of higher order construct is essential, and redundancy analysis was performed. The next essential step was analysis of importance and performance of higher order construct's indicator and performed through following the convergent validity criteria given by Hair et al. (2022).

4.2.1 First Order Measurement Model Assessment

As per the recommendations of Hair et al. (2019, 2022), the first-order and second-order evaluations for measurement model assessments were carried out. Table 8.4 displays indicator reliabilities of all latent constructs followed by internal consistency (Henseler's rhoA and composite reliability) and convergent validity (average variance extracted). By using the HTMT ratio of correlations and Fornell Larker criteria the discriminant validity was examined. The Variance Inflation Factor (VIF) value was used to analyse indicator multicollinearity. Table 8.4 lists the study's first-order construct quality criteria.

4.2.1.1 Factor Loadings. Factor loading range from −1 to +1, higher the loading depicts stronger the relation with the construct. Table 8.5 shows that all the items in the study had factor loadings between ≥0.708 and 0.95 (except BS), recommended by Hair et al. (2019). Hence factor loadings were not an issue in this study (Table 8.4).

Table 8.4. Measurement Model Assessment.

Constructs	Code	Outer loading	VIF	rho_A	CR	AVE
Border Security (BS)	BS1	0.693	1.253	0.791	0.795	0.568
	BS2	0.658	1.254			
	BS3	0.889	1.277			
Event Promotion (EP)	EP1	0.895	2.278	0.837	0.895	0.74
	EP2	0.873	2.343			
	EP3	0.811	1.536			
Infrastructure Development (ID)	ID1	0.772	2.138	0.843	0.888	0.727
	ID2	0.853	1.93			
	ID3	0.927	3.377			
Livelihood Creation (LC)	LC1	0.913	1.964	0.828	0.851	0.66
	LC2	0.857	1.74			
	LC3	0.642	1.296			

Table 8.4. *(Continued)*

Constructs	Code	Outer loading	VIF	rho_A	CR	AVE
Out Migration (OM)	OM1	0.931	1.3	0.79	0.841	0.728
	OM2	0.767	1.3			
Quality of Services (QoS)	QoS1	0.855	1.814	0.803	0.878	0.705
	QoS2	0.802	1.577			
	QoS3	0.861	1.678			
Revisit Intention (RI)	RI1	0.595	1.186	0.848	0.818	0.606
	RI2	0.8	1.534			
	RI3	0.907	1.515			
Local Resource Conservation (LRC)	LRC1	0.929	4.631	0.93	0.891	0.733
	LRC2	0.887	1.996			
	LRC3	0.741	3.043			
Word of Mouth (WoM)	WoM1	0.924	1.991	0.827	0.921	0.853
	WoM2	0.923	1.991			

Source: Researcher.

Table 8.5. Discriminant Validity (Heterotrait-Monotrait Ratio).

	BS	EP	ID	LC	OM	QoS	RI	LRC	WoM
BS									
EP	*0.291*								
ID	0.403	*0.181*							
LC	0.382	0.674	*0.447*						
OM	0.516	0.118	0.818	*0.564*					
QoS	0.647	0.593	0.415	0.684	*0.621*				
RI	0.356	0.273	0.356	0.286	0.349	*0.367*			
LRC	0.336	0.156	0.303	0.26	0.41	0.363	*0.242*		
WoM	0.513	0.34	0.129	0.254	0.118	0.233	0.278	*0.429*	

Source: Researcher.

4.2.1.2 Reliability Analysis. Reliability is the stability and consistency of data collection tool (Mark, 1996). The reliability of construct can be assessed on the bases of CR and AVE (Hair et al., 2019). The indicator BS was retained because it passes the minimum value criteria for AVE and CR (Hair et al., 2019). Furthermore, all CR values presented in Table 8.4 were greater than 0.70 (Hair et al., 2019), indicating that indicators for all constructs were reliable. Hence, construct reliability is established.

4.3 Construct Validity

Construct validity is established when it has discriminant and convergent validity.

4.3.1 Convergent Validity

The convergent validity of a construct is measured by the extent to which it converges to explain the variance of its items (Hair et al., 2019). When the AVE value is less than the recommended value of 0.50, items do not converge to measure the underline construct (Fornell & Larcker, 1981). This AVE value is ≥0.50 (Table 8.4) for all the constructs and explains at least 50% of the variance of its items (Hair et al., 2019). Hence, convergent validity is not an issue.

4.3.2 Discriminant Validity

Discriminant validity is the degree to which measures of different constructs are distinct. The notion is that if two or more concepts are unique, then valid measures of each should not correlate too highly (Bagozzi, 2011). According to Fornell and Larcker's (1981) criterion, discriminant validity is established when the square root of the AVE of a construct is greater than its correlation with all other constructs. In this chapter, the square root of the AVE for a construct was found to be greater than its correlation with other constructs (Table 8.5).

The HTMT ratio of correlations was used to examine the discriminant validity of first-order constructs. According to Henseler et al. (2015), all HTMT ratios were determined to be below the cut off value of 0.85, which represents the proper discriminant validity of all latent variables in the study. Table 8.5 shows that the estimated ratio is less than the required 0.85. As a result, there is strong support for discriminant validity.

4.3.3 Cross Loading

Table 8.6 shows that the highest loading of individual item was loaded highest under the same construct. Hence discriminant validity is not an issue.

4.3.4 Indicator Multicollinearity

VIF statistics is utilised to assess the multicollinearity in the indicators (Fornell & Bookstein, 1982). According to Hair et al. (2019) multicollinearity is not a serious issue if the value for VIF is below 3. The VIF values for the indicators in the study and revealed that VIF for each of the indicators is below the recommended threshold (refer to Table 8.4).

Table 8.6. Discriminant Validity: Cross Loading.

	BS	EP	ID	LC	OM	QoS	RI	LRC	WoM
BS1	**0.693**	0.068	0.064	−0.037	0.184	0.282	−0.075	0.049	0.232
BS2	**0.658**	−0.181	0.358	0.24	0.285	0.307	0.168	0.048	0.146
BS3	**0.889**	0.168	0.165	0.259	0.388	0.503	−0.139	0.462	0.482
EP1	0.064	**0.895**	0.082	0.435	0.047	0.499	−0.012	0.111	0.23
EP2	0.087	**0.873**	−0.004	0.397	0.05	0.313	−0.028	0.096	0.287
EP3	0.061	**0.811**	0.093	0.629	0.052	0.423	−0.105	0.168	0.207
ID1	0.256	0.177	**0.772**	0.229	0.375	0.283	0.089	0.134	0.122
ID2	0.164	−0.063	**0.853**	0.359	0.632	0.303	0.321	0.229	0.066
ID3	0.188	0.118	**0.927**	0.263	0.577	0.25	0.224	0.311	0.084
LC1	0.309	0.576	0.307	**0.913**	0.403	0.498	−0.282	0.209	0.166
LC2	0.065	0.578	0.232	**0.857**	0.282	0.497	−0.136	0.061	−0.133
LC3	0.153	0.079	0.343	**0.642**	0.337	0.303	0.109	0.262	0.187
OM1	0.45	0.1	0.625	0.453	**0.931**	0.479	−0.136	0.347	0.072
OM2	0.16	−0.038	0.432	0.194	**0.767**	0.289	0.061	0.255	0.076
QoS1	0.376	0.433	0.199	0.453	0.301	**0.855**	0.256	0.199	0.274
QoS2	0.292	0.403	0.379	0.388	0.44	**0.802**	0.097	0.319	0.076
QoS3	0.583	0.397	0.262	0.523	0.443	**0.861**	0.096	0.28	0.126
RI1	0.171	0.168	0.181	0.006	0.276	0.368	**0.595**	0.339	0.231
RI2	−0.028	−0.204	0.067	−0.28	−0.117	0.024	**0.8**	−0.078	0.065

(Continued)

Table 8.6. *(Continued)*

	BS	EP	ID	LC	OM	QoS	RI	LRC	WoM
RI3	−0.164	−0.04	0.316	−0.126	−0.166	0.12	**0.907**	0.018	0.035
LRC1	0.347	0.124	0.094	0.103	0.209	0.243	0.077	**0.929**	0.373
LRC2	0.251	0.186	0.455	0.319	0.461	0.35	0.061	**0.887**	0.422
LRC3	0.269	−0.023	−0.056	−0.071	0.157	0.149	0.024	**0.741**	0.131
WoM1	0.383	0.222	0.132	0.077	0.059	0.245	0.223	0.403	**0.924**
WoM2	0.418	0.291	0.056	0.058	0.097	0.104	−0.032	0.344	**0.923**

Source: Researcher.

4.4 Validating Higher Order Construct

The assessment of VVPP in the present study the model was evaluated through the reflective-formative method (Sarstedt et al., 2019). Vibrant village performance was the higher order construct in the study based on six lower order constructs (LOCs). The latent variable scores of LOCs of VVP were applied to measure the higher order construct assessment. Table 8.7 shows that VIF outer values of all formative indicators were below threshold limit of 3 (Hair et al., 2019). It shows that there is no multicollinearity issue in formative assessment of VVPP. The convergent validity was conducted by a single global item (Cheah et al., 2019) to capture the essence of higher order construct of VVPP. According to Chin (1998), if R of global item is 0.7 or more implies it has well explained the convergent validity of HOC. In this study it was 0.689 (Table 8.7) which is less than the suggested one. Hence confidence interval was calculated on 0.95% and it was found significant (Table 8.7). Hence, convergent validity was established as a single global item to capture the essence of higher order construct of VVPP. Besides, the importance and relevance of formative indicators were assessed by 5,000 bootstrapping methods to investigate outer weights. In order to establish the higher order construct validity Outer Weights, Outer Loadings, redundancy analysis of global item and VIF were needed to assess (Hair et al., 2019). The outer weights were found significant for all the six LOCs. Finally, VIF values were assessed to check collinearity. All VIF values (Table 8.8) are less than the recommended value of 3 (Hair et al., 2019). Hence, HOC validity was established.

Table 8.7 shows that LOCs of HOC are independent of each other. These six dimensions are important for VVPP measurement as well as this study according to the respondents. Through redundancy analysis of global item, the standardised beta value was measured, i.e. 0.689 which is not as per the guidelines of Sarstedt et al. (2013). Hence, confidence interval was calculated through smart PLS bootstrapping at 0.95 % and it was found significant and convergent validity of construct (VVPP) was achieved.

4.5 Structural Model Assessment

The outcomes of the structural model's hypothesis testing, as well as the model's explanatory and predictive capacity, were evaluated using the recommendations of Hair et al. (2019, 2020, 2022). The upper value of VIF inner was used less than 3 (Hair et al., 2019) in this study to check the construct multicollinearity. The size and significance of path coefficients in structural model were tabulated in line with Saari et al. (2021). All path coefficients were statistically significant ($p < 0.05$) except (H4) and with large f-square values RI - > WoM, medium f-square values for VVPP - > WoM; RI - > WoM and low f-square value for QoS - > RI (Cohen, 1988) (Table 8.8). Table 8.8 shows that VVP has a significant and positive impact on QoS and WoM. Hence, Hypothesis H1, H2 and H3 are established, while H4 is not accepted.

Table 8.7 Higher Order Construct Validation.

HOC	Formative Indicators	Outer Weight	Outer Loadings	CI (0.95)	Significance? (Weight/ Loading)	VIF (Outer)
VVPP	BS	0.342	0.645	[0.264–0.575]/ [0.229–0.808]	Yes/yes	1.253
	EP	0.296	0.54	[0.147–0.54]/ [0.217–0.818]	Yes/yes	1.588
	ID	0.18	0.592	[0.085–0.296]/ [0.207–0.75]	Yes/yes	1.715
	LC	0.27	0.725	[0.17–0.45]/ [0.397–0.867]	Yes/yes	1.905
	LRC	0.249	0.593	[0.148–0.432]/ [0.159–0.781]	Yes/yes	1.232
	OM	0.24	0.707	[0.094–0.379]/ [0.213–0.872]	Yes/yes	2.132
Redundancy analysis	Indicator	Std. Beta	–	CI (0.95)	Significance	–
VVPP	Global item	0.689	–	[0.402–0.798]	YES	

Source: Researcher.

Table 8.8. Path Coefficients Results and Effect Size.

Relationship	β	*t*-value	*p*-value	CI/Decision	Effect Size (f^2)	VIF Inner	R^2
QoS - > RI (H3)	0.341	2.721	0.007	[0.034–0.527]/ Accepted	0.131 (small)	1.000	
RI - > WoM (H4)	0.136	1.025	0.306	[-0.152-0.4]/ *Rejected*	0.021 (medium)	1.045	
VVPP - > QoS (H1)	0.716	9.986	0	[0.553–0.831]/ Accepted	1.053 (large)	1.000	0.513
VVPP - > WoM (H2)	0.361	2.808	0.005	[0.078–0.585]/ Accepted	0.150 (medium)	1.045	0.169

Source: Researcher.

4.6 PLS Predict

According to Shmueli et al. (2019), PLS predict use manifest variable (MV) summary to compare PLS values and linear regression model (LM) values for all items. In this study, the indicators' MAE values were compared with LM because distribution of prediction errors was non-symmetric (Danks & Ray, 2018; Shmueli et al., 2019). According to Shmueli et al. (2019), if all indicators in default model have lower PLS-SEM value than MAE. The model has high predictive power. Table 8.9 shows that $Q^2 > 0$ and all the indicators have medium PLS-SEM values than MAE, which recognise that model presented in current study had medium predictive relevance.

Table 8.9. PLS Predict.

Construct	Indicators	Q^2 predict	LM_RMSE	LM_MAE	Predictive Relevance
Quality of Services	QoS1	0.223	0.632	0.663	Medium
	QoS2	0.297	0.677	0.727	
	QoS3	0.419	0.595	0.608	
Revisit Intention	RI1	0.073	0.709	0.712	Medium
	RI2	−0.08	0.88	0.891	
	RI3	−0.064	0.79	0.725	
Word of Mouth	WoM1	0.079	0.866	0.86	Medium
	WoM2	0.092	0.577	0.653	

Source: Researcher.

4.7 IPMA

IPMA is important for future research because it provides the importance-performance status of constructs considered in the model. According to Sternad Zabukovšek et al. (2022), IPMA is used in the last part of the research to assess the performance of key factors influencing the key dependent variables of the model. It provides additional results and important information that add value to the PLS-SEM findings. The importance is plotted on the *X* axis, while the performance is plotted on the *Y* axis. Fig. 8.2 shows Accessibility, Activity, Awareness, Availability and Assurance are highly important as well as performing highly and playing an effective role in DRI.

Fig. 8.2 shows that High performance-low importance quadrant have ID, OM, LC, EP, BS; Low performance-low importance quadrant have LRC; Low Performance-High importance quadrant have QoS2, QoS1 and High performance-High importance quadrant have QoS3. It shows that as per the opinion of respondents that majority of the constructs, i.e. ID, OM, LC, EP, BS are less important to measure the VVPP.

4.8 Socio-economic Impacts of Last Run Edition 1

It was the success of the Last Run edition 1, which resulted in the organization of Razang La Full Moon Marathon (UT Union Territory of Ladakh, 2023) in

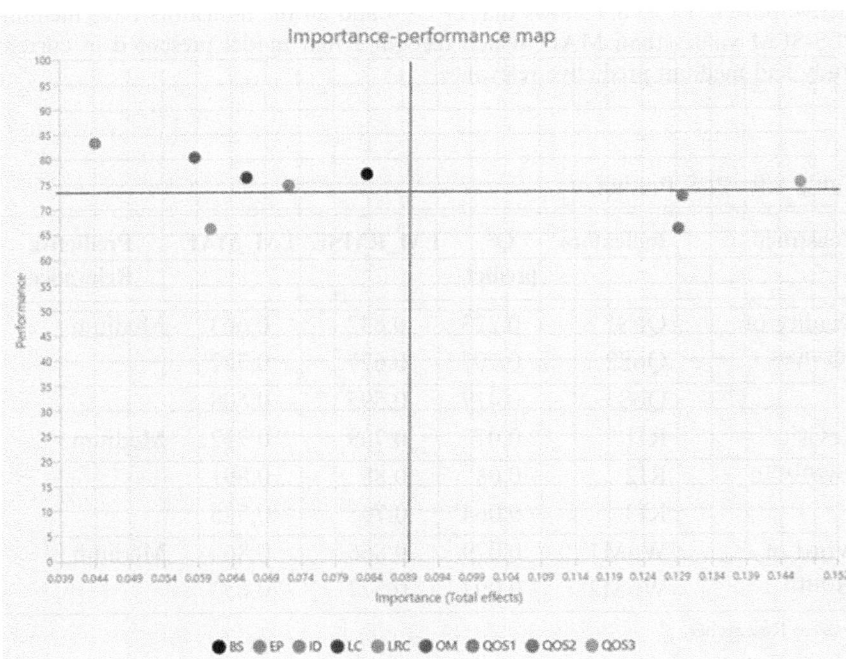

Fig. 8.2. IPMA. *Source:* Researcher.

August 2023 and Last Run edition 2 is scheduled in February 2024 (ASFL, 2023). One of the respondents (organiser) mentioned that

> the Last Run edition 1 was a successful event because it promoted the region at global platform, which motivated us to organize Last Run edition 2. The Last Run edition 2 has already crossed last year's registration count. This year more accommodation properties are open than last year. Also, more residents are staying back in the region than last year, and more cultural events are planned.

The achievement of goals set for Last Run edition 1, motivated the organiser to conduct Last Run edition 2. It was found that economic output as well as social benefits is on the rise year by year. As per Table 8.10, the respondents mentioned that till November 2023, 103 participants have registered which is 37 % higher than Last Run edition 1. During Last Run edition 1, the all-inclusive registration was 558 USD which has increased to 738 USD in Last Run edition 2. It shows 32 % growth in registration price. Also, in edition 2 approximately 25 home stays are expected to be operational, which is 56% more than at the last event. Besides, more local, i.e. approximately 135 are expected to stay in border villages during winter, which is 42% higher than Last Run edition 1. It can be inferred that VVP has visible impact on economy and society on Indo-China border in Pangong region.

Table 8.10. Socio-economic Impacts of Event.

Sr. No.	Parameter	Last Run Edition 1 (February 2023)	Last Run Edition 2 (February 2024)	% Growth
1	Participant registered	75 (confirmed)	103 (till November 2023)	37%
2	Participation fee	558 USD (all-inclusive)	738 USD 252 USD (partial inclusive)	32% (all-inclusive package)
3	Homestay open	16 (average no. of rooms 3)	25 (average no. of rooms 3)	56%
4	Local stayed during winter	95	135 (expected)	42%
5	No. of days for business	4 (Indo-China Border: from Lukung Village to Merak Village)	4 Indo-China Border: from Lukung Village to Merak Village)	–

Source: Researchers.

5. Discussion, Conclusion and Recommendation

Ladakh hosted India's first frozen-lake marathon at Pangong Tso, a lake situated at a height of 13,862 feet. The marathon, known as the 'Last Run', aims to raise awareness about climate change and community development through events at Indo-China border. The event was organised by the Adventure Sports Foundation of Ladakh in collaboration with the Ladakh Autonomous Hill Development Council-Leh, and the Leh district administration. The marathon covered a distance of 21 km, started from Lukung and ended at Maan village. A total of 75 athletes from India and abroad participated, with the opportunity to set a Guinness world record for the world's highest frozen-lake marathon. The Indian Army and the Indo-Tibetan Border Police involved in the implementation of a comprehensive action plan to ensure the success of the event. The marathon's significance lies in promoting sustainable development and a carbon neutral Ladakh. There is need to work on basic amenities in Pangong region. The QoS improvement in Pangong region can enhance the revisit intention of tourists. Also, there is need to promote sports tourism in Ladakh and can be tool for economic development as sports person stay for at least 7 days. Also, there is need to organise more events near border because it improves the infrastructure in border areas. The socio-economic impacts of Last Run edition 1 are clearly visible across the border villages. This event is successful to lower down local migration to Leh city, generating livelihood opportunities, local participation in event and promotion of local culture globally. Hence, the event is economically viable and socially accepted. It can be clearly stated that event organising at border area is promoting border tourism, economic development, de migration, strategic advantage to the Indian security forces and promotion of culture at global platform. Also, this study will strengthen the literature on VVPP measurement and an in-depth understanding of relationship among VVP, WoM, QoS and RI. It was conducted among domestic tourists and future studies can be extended to Indian Himalayan region where VVP is being implemented.

References

Administration of Union Territory of Ladakh (2023, February 23). *Ladakh sets Guinness World Record with Pangong Frozen Lake Marathon.* https://ladakh.gov.in. Accessed on June 8, 2024.

Agarwal, R., & Karahanna, E. (2000). Time flies when you're having fun: Cognitive absorption and beliefs about information technology usage. *MIS Quarterly, 24,* 665–694.

Aguilera, U., Lopez-Novoa, U., Emaldi, M., López-De-Ipiña, D., Pérez-De-Albeniz, I., Valerdi, D., ..., Arza, E. (2016, July). A platform for overcrowding detection in indoor events using scalable technologies. In *2016 Intl IEEE Conferences on Ubiquitous Intelligence & Computing, Advanced and Trusted Computing, Scalable Computing and Communications, Cloud and Big Data Computing, Internet of People, and Smart World Congress (UIC/ATC/ScalCom/CBDCom/IoP/Smart-World)* (pp. 269–276).

Ajzen, I. (1991). The theory of planned behaviour. *Organizational Behavior and Human Decision Processes*, *50*(2), 179–211.

ASFL (2023). *2nd edition of Pangong Frozen Lake Marathon*. ASFL. http://www.asfladakh.com/

Bagozzi, R. P. (2011). Measurement and meaning in information systems and organizational research: Methodological foundations. *MIS Quarterly*, *35*(2), 261–292.

Bandi Tanner, M., Künzi, A., Lehmann Friedli, T., & Müller, H. (2018). Event performance index: A holistic valuation tool. *International Journal of Event and Festival Management*, *9*(2), 166–182.

Becker J.-M., Klein, K., & Wetzels M. (2012). Hierarchical latent variable models in PLS-SEM: Guidelines for using reflective-formative type models. *Long Range Planning*, *45*, 359-394.

Bohlin, M. (2000). Traveling to events. In L. Mossberg (Ed.), *Evaluation of events: Scandinavian experiences* (pp. 13–29). Cognizant.

Bringas, N., & Verduzco, B. (2008). The construction of the northern border as a tourist destination in a context of security alerts. *Region and Society*, *20*(42), 3–36.

Cheah, J. H., Ting, H., Ramayah, T., Memon, M. A., Cham, T. H., & Ciavolino, E. (2019). A comparison of five reflective-formative estimation approaches: Reconsideration and recommendations for tourism research. *Quality and Quantity*, *53*(3), 1421–1458.

Chen, P. J., Singh, D., Bulent Ozturk, A., & Makki, A. (2014). Can fundraising be fun? An event management study of unique experiences, performance and quality. *Tourism Review*, *69*(4), 310–328.

Chin, W. W. (1998). The partial least squares approach to structural equation modeling. In G. A. Marcoulides (Ed.), *Modern methods for business research* (pp. 295–358). Erlbaum.

Cohen, J. (1988). *Statistical power analysis for the behavioral sciences*. Erlbaum.

Connell, J., Page, S. J., & Meyer, D. (2015). Visitor attractions and events: Responding to seasonality. *Tourism Management*, *46*, 283–298.

Crompton, J., & Love, L. (1995). The predictive validity of alternative approaches to evaluating quality of a festival. *Journal of Travel Research*, *34*(1), 11–25.

Cronin, J. J., & Taylor, S. A. (1992). Measuring service quality: A re-examination and extension. *Journal of Marketing*, *56*(3), 55–68.

Danks, N., & Ray, S. (2018). Predictions from partial least squares models. In F. Ali, S. M. Rasoolimanesh, & C. Cobanoglu (Eds.), *Applying partial least squares in tourism and hospitality research*, (pp. 35–52). Emerald.

del Río, J. A. J., Agüera, F. O., Cuadra, S. M., & Morales, P. C. (2017). Satisfaction in border tourism: An analysis with structural equations. *European Research on Management and Business Economics*, *23*(2), 103–112.

ETTravelworld. (2022, June 22). *Border tourism can become a driver of economic upliftment along international border: RD-North India Tourism*. The Economic Times. https://travel.economictimes.indiatimes.com/news/destination/states/border-tourism-can-become-a-driver-of-economic-upliftment-along-international-borders-rd-north-india-tourism/92409003

Fornell, C. G., & Bookstein, F. L. (1982). Two structural equation models: LISREL and PLS applied to consumer exit-voice theory. *Journal of Marketing Research*, *19*(4), 440–452.

Fornell, C. G., & Larcker, D. F. (1981). Evaluating structural equation models with unobservable variables and measurement error. *Journal of Marketing Research, 18*(1), 39–50.

Fragkogianni, M. (2018). Formulating event loyalty in tourism: Lessons from the world travel market. *Event Management, 22*(1), 37–47.

Funchal, B., & Pinto, J. P. (2018). Corporate events' performance and corporate governance: The Brazilian evidence. *Corporate Governance: The International Journal of Business in Society, 18*(1), 14–34.

Gelbman, A., & Timothy, D. J. (2011). Border complexity, tourism and international enclaves: A case study. *Annals of Tourism Research, 38*(1), 110–131.

Getz, D. (2005). *Event management and event tourism* (2nd ed.). Cognizant.

Getz, D. (2008). Event tourism: Definition, evolution, and research. *Tourism Management, 29*(3), 403–428.

Getz, D., Anderson, D., & Sheehan, L. (1998). Roles, issues and strategies for convention and visitors' bureaux in destination planning and product development: A survey of Canadian bureaux. *Tourism Management, 19*(4), 331–340.

Getz, D., Andersson, T., & Larson, M. (2007). Festival stakeholder roles: Concepts and case studies. *Event Management, 10*(2/3), 103–122.

Goodall, S. K. (2004). Rural-to-urban migration and urbanization in Leh, Ladakh. *Mountain Research and Development, 24*(3), 220–227.

Gunn, C. A. (1979). *Tourism planning*. Crane, Russak & Co. Inc.

Hair, J. F., Howard, M. C., & Nitzl, C. (2020). Assessing measurement model quality in PLSSEM using confirmatory composite analysis. *Journal of Business Research, 53*(3), 1421–1458.

Hair, J. F., Hult, G. T. M., Ringle, C. M., & Sarstedt, M. (2022). *A primer on partial least squares structural equation modeling (PLS-SEM)* (3rd ed.). SAGE.

Hair, J. F., Ringle, C. M., & Sarstedt, M. (2011). PLS-SEM: Indeed a silver bullet. *Journal of Marketing Theory and Practice, 19*(2), 139–152.

Hair, J. F., Risher, J. J., Sarstedt, M., & Ringle, C. M. (2019). When to use and how to report the results of PLS-SEM. *European Business Review, 31*(1), 2–24.

Henseler, J., Ringle, C. M., & Sarstedt, M. (2015). A new criterion for assessing discriminant validity in variance-based structural equation modeling. *Journal of the Academy of Marketing Science, 43*(1), 115–135.

Hussain, A. (2022, November 2). Climate migration: Nomads move to towns in warming Ladakh. *The Diplomat: South Asia Edition*. https://thediplomat.com/2022/11/climate-migration-nomads-move-to-towns-in-warming-ladakh/

Janiskee, R. (1996). The temporal distribution of America's community festivals. *Festival Management and Event Tourism, 3*(3), 129–137.

Johnston, R., & Clark, G. (2008). *Service operations and management*. Prentice Hall.

Kim, H. J., Park, J., Kim, M., & Ryu, K. (2013). Does perceived restaurant food healthiness matter? Its influence on value, satisfaction and revisit intentions in restaurant operations in South Korea. *International Journal of Hospitality Management, 33*, 397–405.

Kone, Z. L., Liu, M. Y., Mattoo, A., Ozden, C., & Sharma, S. (2018). Internal borders and migration in India. *Journal of Economic Geography, 18*(4), 729–759.

Kruger, M., & Saayman, M. (2015). Attendance at the U2 concert – Is it a case of 'this is a man's world. *Event Management, 19*(1), 15–32.

Kusonwattana, P., Prasetyo, Y. T., Vincent, S., Christofelix, J., Amudra, A., Montgomery, H. J., ..., Persada, S. F. (2022). Determining factors affecting behavioral intention to organize an online event during the COVID-19 pandemic. *Sustainability*, *14*(20), 12964.

Loi, L. T. I., So, A. S. I., Lo, I. S., & Fong, L. H. N. (2017). Does the quality of tourist shuttles influence revisit intention through destination image and satisfaction? The case of Macao. *Journal of Hospitality and Tourism Management*, *32*, 115–123.

Mark, R. (1996). *Research made simple: A handbook for social workers*. SAGE.

McHone, W. W., & Rungeling, B. (2000). Practical Issues in Measuring the Impact of a Cultural Tourist Event in a Major Tourist Destination. *Journal of Travel Research*, *38*(3), 299–302.

Mendes, J., Do Valle, P. O., & Guerreiro, M. (2013). Destination image and events: A structural model for the Algarve case. In *Event Tourism and Cultural Tourism* (pp. 128–146). Routledge.

Ministry of Home Affairs. (2023a, November 1). Accessed from https://www.mha.gov.in/sites/default/files/BMIntro-1011.pdf

Ministry of Home Affairs. (2023b, August 8). *Vibrant Village Programe*. PIB. https://pib.gov.in/PressReleaseIframePage.aspx?PRID=1946688.

Ministry of Home Affairs. (2023c, July 26). *Vibrant Village Programe*. PIB. https://pib.gov.in/PressReleaseIframePage.aspx?PRID=1942876

Namgail, D., & Chaudhary, R. (2023). The patterns and nature of rural to urban migration in Leh Ladakh. *Asian Resonance*, *12*(1).

Neely, A., Adams, C., & Kennerley, M. (2002). *The performance prism: The scorecard for measuring and managing business success* (p. FT). Prentice-Hall.

Ngoc, K. M., & Trinh, N. T. (2015). Factors affecting tourists' return intention towards Vung Tau city, Vietnam-a mediation analysis of destination satisfaction. *Journal of Advanced Management Science*, *3*(4), 1–7.

Oliver, R. L. (1999). Whence consumer loyalty. *Journal of Marketing*, *63*(4), 33–34.

Özdemir, G., & Çulha, O. (2009). Satisfaction and loyalty of festival visitors. *Anatolia*, *20*(2), 359–373.

Özdemir, A., Tozlu, E., Şen, E., & Ateşoğlu, H. (2016). Analyses of word-of-mouth communication and its effect on students' university preferences. *Procedia-Social and Behavioral Sciences*, *235*, 22–35.

Prabha, P., Rolfe, J., & Sinden, J. (2006). A travel cost analysis of the value of special events: Gem fest in Central Queensland. *Tourism Economics*, *12*(3), 403–420.

Prado-Gascó, V., Calabuig Moreno, F., Ano Sanz, V., Núñez-Pomar, J., & Crespo Hervas, J. (2017). To post or not to post: Social media sharing and sporting event performance. *Psychology and Marketing*, *34*(11), 995–1003.

Quinn, B. (2006). Problematising 'Festival Tourism: Arts Festivals and Sustainable Development in Ireland. *Journal of Sustainable Tourism*, *14*(3), 288–306.

Rachael, E. N., & Douglas, G. P. (2001). Why do people attend events: A comparative analysis of visitor motivations at four South Island events. *Journal of Travel Research*, *39*(4), 449–460.

Ringle, C. M., Sarstedt, M., & Straub, D. W. (2012). Editor's comments: A critical look at the use of PLS-SEM in MIS Quarterly. *MIS Quarterly*, *36*(1), 3–14.

Ringle, C. M., Wende, S., & Becker, J.-M. (2015). *SmartPLS 3, SmartPLS*. Bönningstedt.

Ritchie, J. R. B. (1984). Assessing the impacts of hallmark events: Conceptual and research issues. *Journal of Travel Research*, *23*(1), 2–11.

Rowold, D. J., Benedico, D. P., Garcia-Bertrand, R., Chennakrishnaiah, S., Alfonso-Sanchez, M. A., Gayden, T., & Herrera, R. J. (2016). Ladakh, India: The land of high passes and genetic heterogeneity reveals a confluence of migrations. *European Journal of Human Genetics*, *24*(3), 442–449.

Ryan, C., Smee, A., Murphy, S., & Getz, D. (1998). New Zealand events: A temporal and regional analysis. *Festival Management and Event Tourism*, *5*(1/2), 71–83.

Saari, U. A., Damberg, S., Frömbling, L., & Ringle, C. M. (2021). Sustainable consumption behavior of Europeans: The influence of environmental knowledge and risk perception on environmental concern and behavioral intention. *Ecological Economics*, *189*.

Sarstedt, M., Hair, J. F., Cheah, J.-H., Becker, J.-M., & Ringle, C. M. (2019). How to specify, estimate, and validate higher-order constructs in PLS-SEM. *Australasian Marketing Journal*, *27*(3), 197–211.

Sarstedt, M., Wilczynski, P., & Melewar, T. C. (2013). Measuring reputation in global markets – A comparison of reputation measures' convergent and criterion validities. *Journal of World Business*, *48*(3), 329–339.

Shmueli, G., Sarstedt, M., Hair, J. F., Cheah, J. H., Ting, H., Vaithilingam, S., & Ringle, C. M. (2019). Predictive model assessment in PLS-SEM: guidelines for using PLSpredict. *European Journal of Marketing*, *53*(11), 2322–2347.

Sternad Zabukovšek, S., Bobek, S., Zabukovšek, U., Kalinić, Z., & Tominc, P. (2022). Enhancing PLS-SEM-enabled research with ANN and IPMA: Research study of enterprise resource planning (ERP) systems' acceptance based on the technology acceptance model (TAM). *Mathematics*, *10*(9), 1379.

Stokes, R. (2004). A framework for the analysis of events—Tourism knowledge networks. *Journal of Hospitality and Tourism Management*, *11*(2), 108–123.

Tapia Ladino, M. (2023). Migrations and Borders: Contributions to Understand Mobility in Cross-border Areas. *Journal of borderlands studies*, *38*(3), 441–459.

Tosun, C., Dedeoglu, B. B., & Fyall, A. (2015). Destination service quality, affective image and revisit intention: The moderating role of past experience. *Journal of Destination Marketing & Management*, *4*(4), 222–234.

Union Territory of Ladakh. (2023). ASFL organizes first ever full moon marathon in Ladakh. https://ladakh.nic.in/afsl-organises-first-ever-full-moon-marathon-in-ladakh/

Vezzoli, S. (2021). How do borders influence migration? Insights from open and closed border regimes in the three Guianas. *Comparative Migration Studies*, *9*(1), 1–23.

Walo, M., Bull, A., & Breen, H. (1996). Achieving economic benefits at local events: A case study of a local sports event. *Festival Management and Event Tourism*, *4*(3–4), 95–106.

Wastl-Walter, D., Varadi, M. M., & Veider, F. (2003). Coping with marginality: To stay or to go. *Journal of Ethnic and Migration Studies*, *29*(5), 797–817.

Wetzels, M., Odekerken-Schröder, G., & Van Oppen, C. (2009). Using PLS path modeling for assessing hierarchical construct models: Guidelines and empirical illustration. *MIS Quarterly*, *33*(1), 177–195.

Whitford, M. (2004). Event public policy development in the Northern Sub-Regional Organisation of Councils, Queensland Australia: Rhetoric or realisation? *Journal of Convention & Event Tourism*, *6*(3), 81–99.

Williams-Oerberg, E. (2016). Educational migration among Ladakhi Youth. *Internal Migration in Contemporary India*, 154–179.

Yoon, Y. S., Lee, J. S., & Lee, C. K. (2010). Measuring festival quality and value affecting visitors' satisfaction and loyalty using a structural approach. *International Journal of Hospitality Management*, *29*(2), 335–342.

Yuan, J. J., Morrison, A. M., Cai, L. A., & Linton, S. (2008). A model of wine tourist behaviour: A festival approach. *International Journal of Tourism Research*, *10*(3), 207–219.

Zeithaml, V. A., Berry, L. L., & Parasuraman, A. (1996). The behavioural consequences of service quality. *Journal of Marketing*, *60*(2), 31–46.

Chapter 9

Examining the Impact of Social Media Communication on Visitor's Loyalty: A Case of Sustainable Tourism

Farah S. Choudhary and Najma Khatoon

University of Jammu, India

Abstract

This research paper's goal is to determine whether social media communication has an impact on the loyalty of visitors. The study will explore the correlation between social media activity and visitors' loyalty by examining social media engagement. The research will take the form of surveys and a total of 385 questionnaires given to visitors at various locations of the Ladakh region, asking them about their social media actions and how they relate to loyalty. The data were analysed and tested using various statistical tools such as factor analysis, regression and structural equation modelling. The study's findings will shed important light on the interconnectivity between social media activity, visitor loyalty and the tourism industry. The study's findings show how environmental issues associated with tourism sites may be made aware to visitors through social media marketing and communication campaigns, which may eventually entice them to return.

Keywords: Social media; communication; destination image; loyalty; tourists

1. Introduction

Social media has evolved over the past several years to play a crucial role in the tourism industry, shaping the way tourists explore, engage with and share their experiences. Social media provides a powerful platform for users to discover new destinations exchange travel information and share photographs and stories with friends and followers. Social media platforms offer various features like sharing, liking and commenting that allow users to interact with the content posted by

Sustainable Tourism, Part B, 165–180

doi:10.1108/978-1-83608-326-920241009

their friends and other users. The main aim of the study is to examine the effects of communication and sharing functions of social media platforms on the loyalty of visitors. Loyalty in this context refers to the sustained commitment and preferences exhibited by individuals towards a particular brand, organisation or platform (Bhat & Sharma, 2022). By investing the role of conversation and sharing on social media, this research aims to give information on the fundamental mechanisms that support visitors' loyalty in the digital realm. Social media platforms users with numerous channels for conversation and sharing ranging from comments, direct messaging and public posts to content sharing functionalities such as retweets, likes and shares. These interactive features foster a sense of engagement, facilitating discussions and enabling the dissemination of information across vast networks of individuals. As users engage in conversations, express their opinions and share content, they actively contribute to the formation of online communities and the amplification of digital word of mouth (Bangare et al., 2022). The impact of communication and sharing on loyalty is multifaceted. On one hand, these functions can enhance the overall user experience, enabling individuals to connect with like-minded individuals, seek recommendations and receive social validation for their preferences (Sharma & Bhat, 2023). Such positive interactions can foster a sense of belonging and affinity towards the platform, increasing visitor loyalty. On the other hand negative conversations or experiences shared on social media can have detrimental effects, potentially ending visitor loyalty and causing reputational damage to brands or organisations. To examine the impact of communication and sharing functionality on visitor loyalty, this research aims to shed light on the mechanism through which social media engagement influences visitor loyalty, thereby providing valuable insights for businesses seeking to leverage these platforms effectively. Hoteliers and travel agents are using social media communication to highlight their destinations. Locations that utilise ideal features within their marketing create a lasting impression on the minds of potential tourists. Therefore, the current study is significant as it focuses on how social media communication affects traveller's decisions and perceptions of places.

2. Theoretical Background

2.1 Social Media

In today's contemporary world, social media is an indispensable component that has revolutionised how people engage and communicate with one another. It has transformed interpersonal interactions and communication, making it a vital element of contemporary life (Khatoon & Choudhary, 2023). Furthermore, social media allows users to participate in communication and engagement activities with one another. They can help and be helped by other users as well as reply to and respond to queries from other users. This is especially true with social media, which is one of the platforms where contact and communication are most deeply felt (Gündüz, 2017). Moreover, travellers today prefer to use search engines and social media platforms to find the information they need, rather than relying on established tour operators or travel companies (Ofitserova, 2021). Further, it has become all the more important for both the researcher and practitioner to study and analyse digital communication and its effect on consumer behaviour, especially in tourism industry to understand the impact of social media in present digital marketing campaigns (Ibrahim & Ganeshbabu, 2018). In a connected world, the travel and tourism sectors have been spending money on social media platforms to increase their advertising and marketing efforts (Zhou & Wang, 2014). Furthermore, besides being the backbone of the world economy, tourism also boosts growth of a nation and helps them to increase and amplify foreign exchange earnings (Trun & Rudolf, 2022). Also, hoteliers and travel agent shave enhanced the visual appearance and content of their websites in order to make tourist destination more recognisable and attract more visitors, new inhabitants, investors and other stakeholders (Kumar & Panda, 2019).

The present study is based on the two most important concepts of social media communication, i.e. how travel agents and hoteliers use their content and reviews of tourists to build their brand (Shu & Chuang, 2011). Moreover, the study also revolves around the importance of *'likes'* on the story and pictures posted on the social media page for any organisation (Kietzmann et al., 2011; Machado et al., 2016).

2.2 Social Media Communication

The social media features show how frequently a user interacts with other people. Most social networking platforms are there to facilitate dialogues between people or groups. Discussions can take place for a variety of reasons, such as for people to connect with like-minded individuals, fall in love, meet new people, increase their self-confidence or be at the forefront of cutting-edge ideas or pressing concerns. Additionally, social media sites let users spread messages that support the community, the environment or charitable organisations. There are online forums where economic and political issues can be discussed. Moreover, sharing videos, photos and stories through social media is also another form of communication (Kietzmann et al., 2011; Machado et al., 2016; Shu & Chuang, 2011).

Furthermore, the extent to which users trade, provide and receive material is determined by this feature component. The term 'social' is described as regularly emphasising the value of one-on-one interactions. But frequently, sociality is about the things that make these intimate connections possible. Additionally, disseminating scientific ideas via social media can foster the development of critical thinking, multipurpose writing and the discipline required to post often (Khatoon & Choudhary, 2024). Kietzmann et al. (2011) proposed two implications that were brought about by sharing, the first being that we must first identify the social objects are those that let their users exchange information or learn about new items that could relate to their shared interests, and they also allow or encourage sharing to some extent. Also, Machado et al. (2016) suggested that for any organisation, the quantity of social media 'shares' is just as significant as likes and comments.

3. Visitors' Loyalty

Studies on the subject have been conducted for decades, and they have shown that visitor loyalty is a key predictor of successful tourism locations. As compared to the costs involved with advertising, loyal visitors stay longer, engage more in a variety of social and cultural events and spread the word favourably about the location (Lau & McKercher, 2004; Molina et al., 2010; Lehto et al., 2004). Furthermore, the term 'loyalty' refers to a favourable attitude towards a good or service, followed by recurrent purchases and word-of-mouth advertising (Stylidis et al., 2020). The idea of loyalty has been studied as one of the most important concept in marketing literature and a vital component for business growth and success (Bauer et al., 2002; La Barbara & Mazursky, 1983; Pine et al., 1995; Turnbull & Wilson, 1989). Operationally, 'destination loyalty' is the extent to which travellers consider a place to be 'recommendable' (Patwardhan et al., 2020) or the degree to which a traveller intends to return to the location (Oppermann, 2000). Moreover, according to Petrick (2004), repeat customers may be less price-sensitive than first-time customers. Besides, the loyalty of visitors is something that travel destinations are increasingly more concerned with (Baloglu et al., 2017) than increasing the number of yearly tourists (Anton et al., 2017) because it provides a consistent stream of revenue from clients that aren't price-sensitive (Woyo & Slabbert, 2020). The concept of loyalty has increased its importance even in the literature of tourism due to its effect on visitors and how it helps the destinations to gain profits (Baloglu, 2001). There are various studies that have examined the relationship of tourism and loyalty (Oppermann, 2000; Yoon & Uysal, 2005). Furthermore, Kim (2018) investigated loyalty and tourism together and suggested that loyalty in tourism is the likelihood and probability of the visitors returning to the location and recommending it to friends and family (Kim, 2018). When it comes to potential tourists, positive word-of-mouth is a reliable source of information (Yoon & Uysal, 2005), especially helpful in the tourist sector, which depends on previous travellers' reviews (Zhang et al., 2014). Another sign of effective destination development is the intention to return, which boosts the competitiveness of tourism locations (Chen & Phou, 2013).

4. Social Media Communication and Visitor's Loyalty

The phrase 'destination loyalty' refers to a concept that illustrates the long-term return behaviour of the tourists towards a location that is determined by their experience at the same destination in the past (Oppermann, 2000). According to Chi and Qu (2008) and Phillips et al. (2013), two behavioural intents that are frequently employed to gauge destination loyalty are the intention to return and good word of mouth. Moreover, Prior research prioritised visitors' intentions to return to the same location, the same category of location or a location within the same area when assessing visitors' intentions to revisit a location (Correia Loureiro & Kaufmann, 2012). A place might be seen as a commodity and fulfil traveller demands, or devoted visitors may return or suggest it to others (Chen et al., 2023). Furthermore, Chen and Tsai (2007) noted that travellers often select a tourist site to visit and undertake post-visit reviews, which will result in future destination loyalty. Also, it is frequently

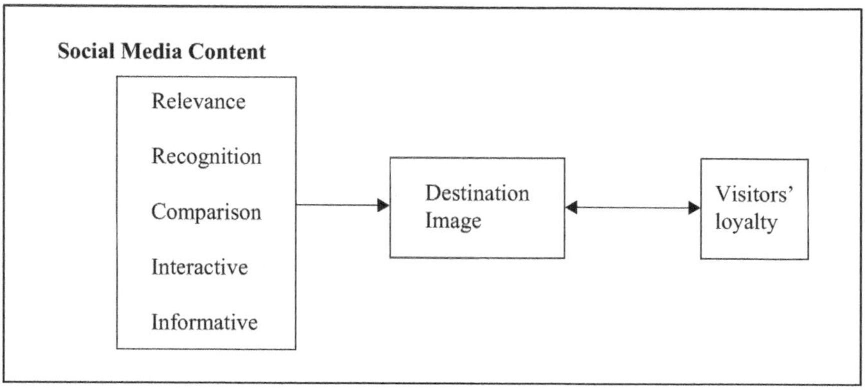

Fig. 9.1. Proposed Model. *Source:* Prepared by author.

believed that satisfying, pleasurable and memorable experiences have an impact on tourist satisfaction and destination loyalty (Jiang et al., 2022). According to Barnes et al. (2016), the largest influences on destination loyalty come from long-term memory events. Furthermore, Pan et al. (2021) studied how different information sources and the methods via which they are distributed influenced how tourists perceive their destination. It was discovered that both natural (like friends and family) and autonomous (like news, documentaries, movies and guidebooks) sources had substantial influence. The most reliable and accurate method for disseminating this knowledge was word of mouth. Furthermore, depending on the sort of location and how well-versed visitors are in it, different tourists will be more or less eager to spread good word of mouth about it (Phillips et al., 2013). Moreover, contrary to conventional means of communication, new technology enables travellers to participate in more vibrant, engaging and modern electronic word of mouth. Based on the above literature, we can hypothesise that (Fig. 9.1),

H1. There exists a positive relationship between social media communication and visitor loyalty.

5. Social Media Communication, Visitor's Loyalty and Destination Image

Loyal visitors are known to spend more time at the location and participate in a variety of activities. They may also encourage others to visit the area by making favourable suggestions (Lehto et al., 2004). San Martín et al. (2019) investigated that visitors' loyalty is a concept that explains the intend to return or refer to friends and family. Furthermore, destination loyalty can be described in the best way as an intention of the visitors to return to the destination and their desire to share positive word of mouth (WOM) with family, relatives and/or friends (Kim, 2018; Lv & McCabe, 2020). Moreover, according to Kislali et al. (2020), loyalty may also manifest as a propensity to talk about a place on social media. Also, social media platforms enable user collaboration, communication and content sharing (Palmer & Koenig-Lewis, 2009). It allows visitors to modify their expectations according to the information they get, independent of the material's author or the person who shares it. They will sculpt their impression of the location if they have complete faith in the publisher's source (Narangajavana et al., 2017). Therefore, tourist locations, particularly those with political barriers, should grasp how social media works, understand users, take charge of positioning and branding and have oversight over the competence of information (Farhangi & Alipour, 2021). There are numerous studies that signify a positive relationship between destination image and loyalty (Akroush et al., 2016; Jeong & Kim, 2019). Stepaniuk (2015) suggested that hoteliers ad travel agents use various social media platforms to post various images, videos, reels and other destination-related information to motivate visitors and stimulate positive destination image among them, thereby encouraging them to revisit the destination which further leads to loyalty towards the destination.

Drawing from this we hypothesise that (Fig. 9.1),

H2. The relationship between social media communication and visitor's loyalty is mediated by destination image.

Fig. 9.1: On the basis of hypothesis developed, Fig. 9.1 has been proposed.

6. Research Methodology

6.1 Measurement Scale

The current study is based on the variables evaluated and measured using five-point Likert scale, ranging from 1 (strongly disagree) to 5 (strongly agree). The conversation and sharing functions of the scale employed in the current research is based on the scale developed and used in past studies of Kietzmann et al. (2011), Silva et al. (2020), Baccarella et al. (2018) and Rainoldi et al. (2018). Additionally, the scale for visitor loyalty is extracted from studies namely Schlesinger et al. (2020), Chen and Rahman (2018) and Mirzaalian and Halpenny (2021). Further, the scale was redesigned and redeveloped according to the respondents and hotel industry of the Ladakh region.

6.2 Data Collection

The data of the current research were gathered by survey approach and using purposive sampling from tourists visiting different locations in the Ladakh region. A total of 385 questionnaires were sent conveniently, and 350 of them were collected and deemed appropriate for further analysis and consideration, representing an 85% response rate. In Ladakh, the tourist industry is one of the most important sectors. Furthermore, Ladakh is the most well-known northern state in the world for its breathtaking natural beauty, amazing landmarks, monasteries, culture, kind people, etc. Ladakh has experienced a notable increase in tourist visits in tandem with the growing trend of data and information exchange. The tourism sector plays a significant role in Ladakh's economy by providing benefits including jobs, foreign exchange, infrastructure development and the expansion of local industries. It is considered to be a significant sustainable tourist destination of India with more than half of the land of Leh covered under protected area network. It is an important international tourist destination where its unique climate, rich culture and heritage make it one of the best sustainable tourist destination of India.

The demographic representation of the sample is presented in Table 9.1.

6.3 Data Analysis

The study employed factor analysis (FA) to streamline and condense the data. The scale was tested using the Cronbach's alpha test for reliability. Using a varimax rotation technique in conjunction with principal component analysis, FA was executed to summarise the initial data into the least and most significant factors. For

Table 9.1. Individual Characteristics.

Variable	Frequency	Percentage
Gender		
Male	198	54.0
Female	152	46.0
Total	350	100.0
Age		
Up to 20 years	43	13.2
21–35 years	191	54.6
36–50 years	105	24.7
Above 50	11	7.5
Total	350	100.0
Nationality		
Indian	227	80.8
Foreigner	123	19.2
Total	350	100.0
Personal Annual Income Level		
Up to 2′00′000	180	35.4
2,00,000–5,00,000	105	25.2
5,00,000–10,00,000	40	22.3
10,00,000 and above	25	17.1
Total	350	100.0

Source: Prepared by author.

additional data analysis, the items of the questionnaire with factor loading less than 0.5 and Eigen value less than 1.0 were disregarded. Furthermore, Kaiser–Meyer–Olkin (KMO) and Barlett's test were conducted. In addition, the KMO for this investigation was 0.784, which is below the permissible threshold of 0.7. The instrument was approved as a result. Following that, a Barlett's test was performed, and the result was 0.000, indicating that FA could once more be performed. Further, to confirm the factor, confirmatory factor analysis (CFA) was applied.

6.4 Stage 1 Factor Analysis (FA)

The 15-item conversation and sharing function of the social media survey's underlying factor structure has been determined using principal component analysis and oblique rotation. The 15-item item conversation and sharing function of the social media instrument's underlying three components, which accounted for more than 69% of the scale data's variance, was established (Table 9.2).

6.5 Stage 2 Confirmatory Factor Analysis (CFA)

In order to confirm the factors, CFA was performed using all the statements of the two factors of social media communication loaded to determine whether the three-factor model was the best and most appropriate conceptualisation of social

Table 9.2. Factor Analysis.

Sharing and Conversation Functions of Social Media					
Dimensions	**Loadings**	**Com.**	**EV**	**VE**	**KMO**
Sharing function3	0.824	0.684	11.376	14.105	0.869
Sharing function2	0.704	0.656			
Sharing function1	0.727	0.818			
Sharing function5	0.772	0.774			
Sharing function4	0.687	0.644			
Sharing function6	0.798	0.773			
Communication function3	0.842	0.809	1.216	13.760	
Communication function1	0.754	0.658			
Communication function2	0.782	0.731			
Visitors Loyalty					
Dimensions	**Loadings**	**Com.**	**EV**	**VE**	**KMO**
Loyalty1	0.722	0.623	1.324	13.780	0.860
Loyalty2	0.864	0.776			
Loyalty3	0.757	0.649			

Source: Result from SPSS.

media communication based on the aforementioned findings. The factor loading of each of the item suggests that as the factor loading of each item is greater than 0.5, thus all the items are retained.

7. Reliability and Validity

The construct reliability and Cronbach's alpha tests used to determine the data's reliability both returned values of more than 0.70, demonstrating the data's dependability (Table 9.5). Additionally, convergent validity and discriminant validity were used to investigate construct validity. The average variance extracted values and factor loadings for all scales are greater than 0.70 and 0.50, respectively. This demonstrated the scale measures' satisfactory convergent validity (Table 9.5). As long as the correlation between distinct scales is greater than the square root of the average variance retrieved for all scales, discriminant validity has been demonstrated. These criteria were satisfied in our situation, and the discriminant validity was deemed adequate. GFI, CFI and NFI goodness-of-fit indices, as well as RMSEA badness-of-fit criteria, were all determined to be more than 0.90 or almost at the limit, respectively (Table 9.4).

7.1 Composite Reliability

In order to evaluate the internal consistency of the data, composite reliability was tested, and the findings revealed that all the constructions sharing (0.842), communication (0.921) and visitor loyalty (0.867) were higher than the standard of 0.70 (Table 9.2) as shown by Hair et al. (2009), and all scale structures have an average value above the squared.

Table 9.3. Measurement Model Fit Indices.

Measures	**$\chi 2$/df**	**Goodness of Fit Index (GFI)**	**Normed Fit Index (NFI)**	**Comparative Fit Index (CFI)**	**Root Mean Square Error of Approximation (RMSEA)**
Sharing	2.131	0.937	0.933	0.952	0.054
Communication	2.335	0.935	0.956	0.943	0.038
Visitors loyalty	2.243	0.954	0.947	0.923	0.048
Overall model fit	3.011	0.948	0.939	0.931	0.041

Source: Result from SPSS.

Table 9.4. Convergent and Discriminant Validity.

Measures	Sharing	Conversation	Visitors Loyalty
Sharing	0.714		
Conversation	(0.421)	0.746	
Visitors loyalty	(0.453)	(534)	0.729

Source: Result from SPSS.

7.2 Convergent and Discriminant Validity

The validity of the scale is tested by looking at its convergent and discriminant validity. The standardized regression weight (SRW) values of every item are discovered to be higher than the 0.50 threshold limit, indicating convergent validity (Table 9.2). Additionally, discriminant validity was evaluated by contrasting the retrieved variance of all the components with the squared, correlation estimates showing the differences between the scales (Table 9.4).

7.3 Hypothesis Testing Using Structural Equation Modelling (SEM)

SEM was used to verify the measurement models and to test the hypothesis. The model's total fitness increased significantly with$\chi 2$/df (3.011), GFI (9.48), NFI (0.939), CFI (0.931) and RMSEA (0.41) (Table 9.3). The investigation showed a strong theoretical connection between conversation and visitors loyalty (SRW = 0.746, $p = 0.000$) showing how effective a visitor can be when his needs are taken into account while building the services. Results showed that sharing is significant and favourable (SRW = 0.521, $P = 0.000$) influencing visitor loyalty (Table 9.5). So, we accept the second hypothesis.

8. Implications

8.1 Academic Contribution

The study shows the importance of social media communication as one of the key motivating factors to create positive destination image, thereby creating loyalty amongst the visitors. The study's findings show how environmental issues associated

Table 9.5. Hypotheses Result (Direct Effect).

Parameters	SRW(β)	*p* Value	Hypothesis	Conclusion
Conversation→ visitors loyalty	0.746	***	*H1*	Supported
Sharing→ visitors loyalty	0.521	***	*H2*	Supported

Source: Result from SPSS.

Note: ***$p < 0.001$.

with tourism sites may be made aware to visitors through social media marketing and communication campaigns, which may eventually entice them to return. The tourist sector benefits from new avenues that social media information sources help open up, enabling them to better investigate and capitalise on the destination's image. Congruent with the past studies, like Yoon and Uysal (2005) and McKercher and Prideaux (2014), a destination's ability to successfully attract tourists depends largely on the content organisations share on their social media platforms. Furthermore, Kim et al. (2017) suggested that when tourists choose to travel, they often find it challenging to determine if a location is nice or not if they haven't been there before. They trust social media posts to develop an image about a destination. Additionally, Xiang and Gretzel (2010) also suggest that social media platforms are utilised in this situation to reduce decision-making ambiguity. Besides, in order to encourage user/visitor activity and create a positive perception of the destination in the minds of users/visitors, tourism providers, according to Stepaniuk (2015), utilise social media platforms to publish photographs and other relevant information about the location. According to Jeong and Kim's (2019) comparable study, visitors are more likely to return or promote a site to others if they have a positive impression of it. Also, numerous studies have found that information from social media platforms increases customer satisfaction, loyalty and inclination to travel to a certain location (Kim, 2018; Oppermann, 2000; Yoon & Uysal, 2005; Zhang et al., 2014). The current study also validates and broadens the conclusion and results of these studies and identifies social media communication as an important factor to influence visitors' loyalty.

8.2 Managerial Contribution

The findings of the study give suggestions and contribute to tourism industry as a whole. The reels, status, videos, images, product evaluations and blog posted and shared by the travel agents and hoteliers create a positive impact on the mind of the visitors and help them to revisit a destination. For the tourist industry to succeed, it is essential to understand the common interests of both existing and future guests. On numerous social media channels, customers will discuss brands, seek help and share evaluations of their encounters with hospitality companies. Moreover, over the last 10 years, Ladakh has seen a significant growth in the number of visitors. Although there is no denying that domestic tourism has increased dramatically since 2010, there is compelling evidence to suggest that this growth has accelerated since 2016 and will continue to do so. According to the Tourism Department of Leh-Ladakh's annual report for 2019, the number of foreign travellers has been increasing steadily although at a rather modest pace since 2004. Social media communication plays a significant part in boosting travel to the Ladakh region Travel agents and hoteliers should leverage social media communication to promote Ladakh as a sustainable green destination. According to the findings of the study, it is clear that more emphasis should be given to establishing strong online communication. In Ladakh, some resorts like Ashuna Eco Resort, Rangyul Resort and Lchangnang Retreat share interesting and attractive photos and videos of their destinations on social media pages showing culture, cuisine, seasonal reflection and facilities to attract and engage tourists and promote sustainable tourism. These hoteliers and travel agents share photos, reels, stories and various information related to their destination on their social media accounts which will be very helpful for any tourists to make it easy to find any destination and create a positive image about the destination in their minds. Social media communication will help to attract new tourists as well as retain the present tourists and gain their loyalty.

9. Conclusion and Future Research

Digital technology has undergone a revolution thanks to social media, which is also a technical marvel. Social media websites include blogs, Google, YouTube, Facebook, Twitter, Instagram and Google+. Online communities and groups can establish and exchange information, ideas, viewpoints and interests. The performance of the tourist business depends heavily on social media. In order to increase the quality of attractions and other services for both current and future tourists, the majority of locations have looked for creative methods to do it. By illuminating the process through which social media participation affects customer loyalty, this study hopes to offer useful information to businesses looking to make the most of these platforms. Exploratory factor analysis (EFA) was used to purify the data, and CFA was used to validate the factorial structure. SEM was employed for the testing of hypotheses. It was employed to determine how much the communication and sharing function of social media is affecting visitors' loyalty. The impact of social media communication on visitors can be witnessed as quite significant. Destination image proved to be a reliable mediator contributing to the relationship between communication/sharing functionality and visitor loyalty. Further it can be fruitful to understand the possible advantages and difficulties of using social media, as well as the approaches and best practices that can be used to foster positive and sustained visitor loyalty. Despite giving some positive outcomes the possibility of limitations can not be ruled out. Like the population of this study was limited to only visitors of a tourist destination in the Ladakh region. Their interactions, reputation, and allegiance may be different from those of travelling to other nations. Therefore, the results from the study may not be generalized beyond this population. To make these results more broadly applicable, it would be essential to conduct comparable research in other popular tourist locations.

Acknowledgement

This paper is an outcome of Research and Seed Grant provided to the Faculty by University of Jammu.

References

Akroush, M. N., Jraisat, L. E., Kurdieh, D. J., AL-Faouri, R. N., & Qatu, L. T. (2016). Tourism service quality and destination loyalty–the mediating role of destination image from international tourists' perspectives. *Tourism Review*.

Anton, C., Camarero, C., & Laguna-Garcia, M. (2017). Towards a new approach of destination loyalty drivers: Satisfaction, visit intensity and tourist motivations. *Current Issues in Tourism*, *20*(3), 238–260.

Baccarella, C. V., Wagner, T. F., Kietzmann, J. H., & McCarthy, I. P. (2018). Social media? It's serious! Understanding the dark side of social media. *European Management Journal*, *36*(4), 431–438.

Baloglu, S. (2001). Image variations of Turkey by familiarity index: Informational and experiential dimensions. *Tourism management*, *22*(2), 127–133

Baloglu, S., Zhong, Y. Y., & Tanford, S. (2017). Casino loyalty: The influence of loyalty program, switching costs, and trust. *Journal of Hospitality & Tourism Research*, *41*(7), 846–868.

Bangare, M. L., Bangare, P. M., Ramirez-Asis, E., & Bhat, D. A. R. (2022). Role of machine learning in improving tourism and education sector. *Materials Today: Proceedings*, *51*, 2457–2461. https://doi.org/10.1016/j.matpr.2021.11.615

Barnes, S. J., Mattsson, J., & Sørensen, F. (2016). Remembered experiences and revisit intentions: A longitudinal study of safari park visitors. *Tourism Management*, *57*, 286–294.

Bauer, H., Mark, G., & Leach, M. (2002). Building customer relations over the Internet. *Industrial Marketing Management*, *31*(2), 155–163.

Bhat, D. A. R., & Sharma, V. (2022). Enabling service innovation and firm performance: The role of co-creation and technological innovation in the hospitality industry. Technology Analysis & Strategic Management, *34*(7), 774–786. https://doi.org/10.1080/09537325.2021.1919614

Chen, J., Hsu, F. C., Yan, L., Lee, H. M., & Zhang, Y. (2023). Tourists' food involvement, place attachment, and destination loyalty: The moderating role of lifestyle. *Behavioral Sciences*, *13*(8), 629.

Chen, C. F., & Phou, S. (2013). A closer look at destination: Image, personality, relationship and loyalty. *Tourism Management*, *36*, 269–278.

Chen, H., & Rahman, I. (2018). Cultural tourism: An analysis of engagement, cultural contact, memorable tourism experience and destination loyalty. *Tourism management perspectives*, *26*, 153–163. https://doi.org/10.1016/j.tmp.2017.10.006

Chen, C. F., & Tsai, D. (2007). How destination image and evaluative factors affect behavioral intentions? *Tourism Management*, *28*(4), 1115–1122.

Chi, C. G. Q., & Qu, H. (2008). Examining the structural relationships of destination image, tourist satisfaction and destination loyalty: An integrated approach. *Tourism Management*, *29*(4), 624–636.

Correia Loureiro, S. M., & Kaufmann, H. R. (2012). Explaining love of wine brands. *Journal of Promotion Management*, *18*(3), 329–343.

Farhangi, S., & Alipour, H. (2021). Social Media as a Catalyst for the Enhancement of Destination Image: Evidence from a Mediterranean Destination with Political Conflict. *Sustainability*, *13*(13), 7276.

Gündüz, U. (2017). The effect of social media on identity construction. *Mediterranean Journal of Social Sciences*, *8*(5), 85.

Hair, J. F., Black, W. C., Babin, J. B., Anderson, R. E., & Tatham, R. L. (2009). *Multivariate Data Analysis* (6th ed.). Pearson Education Inc.

Ibrahim, S. S., & Ganeshbabu, P. (2018). A study on the impact of social media marketing trends on digital marketing. *Shanlax International Journal of Management*, *6*(1), 120–125.

Jeong, Y., & Kim, S. (2019). A study of event quality, destination image, perceived value, tourist satisfaction, and destination loyalty among sport tourists. *Asia Pacific Journal of Marketing and Logistics*, *32*(4), 940–960.

Jiang, L., Eck, T., & An, S. (2022). A study on the effect of emotional solidarity on memorable tourism experience and destination loyalty in volunteer tourism. *Sage Open*, *12*(1), https://doi.org/10.1177/21582440221087263

Khatoon, N., & Choudhary, F. (2023). A study on functional building blocks of social media–A bibliometric analysis. *European Economic Letters (EEL)*, *13*(4), 1–8.

Khatoon, N., & Choudhary, F. (2024). *Strategizing green destination image through social media functionality: A study of the tourism industry*. Business Strategy & Development.

Kietzmann, J. H., Hermkens, K., McCarthy, I. P., & Silvestre, B. S. (2011). Social media? Get serious! Understanding the functional building blocks of social media. *Business Horizons*, *54*(3), 241–251.

Kim, J. H. (2018). The impact of memorable tourism experiences on loyalty behaviors: The mediating effects of destination image and satisfaction. *Journal of Travel Research*, *57*(7), 856–870.

Kim, S. E., Lee, K. Y., Shin, S. I., & Yang, S. B. (2017). Effects of tourism information quality in social media on destination image formation: The case of Sina Weibo. *Information and Management*, *54*(6), 687–702.

Kislali, H., Kavaratzis, M., & Saren, M. (2020). Destination image formation: Towards a holistic approach. *International Journal of Tourism Research*, *22*(2), 266–276.

Kumar, N., & Panda, R. K. (2019). Place branding and place marketing: A contemporary analysis of the literature and usage of terminology. *International Review on Public and Nonprofit Marketing*, *16*, 255–292.

La Barbara, P. A., & Mazursky, D. (1983). A longitudinal assessment of consumer satisfaction/dissatisfaction: The dynamic aspect of the cognitive process. *Journal of Marketing Research*, *20*, 393–404.

Lau, L., & McKercher, B. (2004). Exploration versus consumption: A comparison of first-time and repeat tourists. *Journal of Travel Research*, *42*, 279–285.

Lehto, X. Y., O'Leary, J. T., & Morrison, A. M. (2004). The Effect of Prior Experience on Vacation Behavior. *Annals of tourism research*, *31*(4), 801–818.

Lv, X., & McCabe, S. (2020). Expanding theory of tourists' destination loyalty: The role of sensory impressions. *Tourism Management*, *77*, 104026.

Machado, J., Azar, S., Vacas de Carvalho, L., & Mender, A. (2016). Motivations to interact with brands on Facebook – Towards a typology of consumer–brand interactions. *Journal of Brand Management*, *23*(2), 153–178.

McKercher, B., & Prideaux, B. (2014). Academic myths of tourism. *Annals of Tourism Research*, *46*, 16–28.

Mirzaalian, F., & Halpenny, E. (2021). Exploring destination loyalty: Application of social media analytics in a nature-based tourism setting. *Journal of Destination Marketing & Management*, *20*(2), 1–10.

Molina, A., Gómez, M., & Martín-Consuegra, D. (2010). Tourism marketing information and destination image management. *African journal of Business management*, *4*(5), 722.

Narangajavana, Y., Fiol, L. J. C., Tena, M. Á. M., Artola, R. M. R., & García, J. S. (2017). The influence of social media in creating expectations. An empirical study for a tourist destination. *Annals of Tourism Research*, *65*, 60–70.

Ofitserova, E. (2021). *The impact of travel content in social media on customers' tourism decisions in Russia*. Master's thesis, Helsinki University. Theseus. https://www.theseus.fi/handle/10024/512250

Oppermann, M. (2000). Tourism destination loyalty. *Journal of Travel Research*, *39*(1), 78–84.

Palmer, A., & Koenig-Lewis, N. (2009). An experiential, social network-based approach to direct marketing. *Direct Marketing: An International Journal*, *3*(3), 162–176.

Pan, X., Rasouli, S., & Timmermans, H. (2021). Investigating tourist destination choice: Effect of destination image from social network members. *Tourism Management*, *83*, 104217.

Patwardhan, V., Ribeiro, M. A., Payini, V., Woosnam, K. M., Mallya, J., & Gopalakrishnan, P. (2020). Visitors' place attachment and destination loyalty: Examining the roles of emotional solidarity and perceived safety. *Journal of Travel Research*, *59*(1), 3–21.

Petrick, J. F. (2004). Are loyal visitors desired visitors? *Tourism Management*, *25*(4), 463–470.

Phillips, W. J., Wolfe, K., Hodur, N., & Leistritz, F. L. (2013). Tourist word of mouth and revisit intentions to rural tourism destinations: A case of North Dakota, USA. *International Journal of Tourism Research*, *15*(1), 93–104.

Pine, B. J., Peppers, D., & Rogers, M. (1995, March–April). Do you want to keep your customers forever? *Harvard Business Review*, 103–114.

Rainoldi, M., Driescher, V., Lisnevska, A., Zvereva, D., Stavinska, A., Relota, J., & Egger, R. (2018). Virtual reality: An innovative tool in destinations' marketing. *The Gaze: Journal of Tourism and Hospitality*, *9*, 53–68.

San Martín, H., Herrero, A., & García de Los Salmones, M. D. M. (2019). An integrative model of destination brand equity and tourist satisfaction. *Current Issues in Tourism*, *22*(16), 1992–2013.

Schlesinger, W., Cervera-Taulet, A., & Pérez-Cabañero, C. (2020). Exploring the links between destination attributes, quality of service experience and loyalty in emerging Mediterranean destinations. *Tourism Management Perspectives*, *35*, 100699.

Sharma, V., & Bhat, D. A. R. (2023). The role of community involvement in sustainable tourism strategies: A social and environmental innovation perspective. *Business Strategy and Development*, *6*, 119–127. https://doi.org/10.1002/bsd2.227

Shu, W., & Chuang, Y. H. (2011). The perceived benefits of six-degree-separation social networks. *Internet Research*, *21*(1), 26–45.

Silva, S. C., Feitosa, W., Duarte, P., & Vasconcelos, M. (2020). How to increase engagement on social media using the honeycomb model: A case study in a Portuguese HR company. *Revista de Gestão*, *27*(2), 153–167.

Stepaniuk, K. (2015). The relation between destination image and social media user engagement – Theoretical approach. *Procedia – Social and Behavioral Sciences*, *213*, 616–621.

Stylidis, D., Woosnam, K. M., Ivkov, M., & Kim, S. S. (2020). Destination loyalty explained through place attachment, destination familiarity and destination image. *International Journal of Tourism Research*, *22*(5), 604–616.

Tran, N. L., & Rudolf, W. (2022). Social media and destination branding in tourism: A systematic review of the literature. *Sustainability*, *14*(20), 13528.

Turnbull, P., & Wilson, D. T. (1989). Developing and protecting profitable customer relationships. *Industrial Marketing Management*, *18*, 233–238.

Woyo, E., & Slabbert, E. (2020). Unpacking the motivations, satisfaction and loyalty of tourists travelling to a distressed destination. *Anatolia*, *31*(4), 536–548.

Xiang, Z., & Gretzel, U. (2010). Role of social media in online travel information search. *Tourism Management*, *31*(2), 179–188.

Yoon, Y., & Uysal, M. (2005). An examination of the effects of motivation and satisfaction on destination loyalty: A structural model. *Tourism Management*, *26*, 45–56.

Zhang, H., Fu, X., Cai, L. A., & Lu, L. (2014). Destination image and tourist loyalty: A meta-analysis. *Tourism Management*, *40*, 213–223.

Zhou, L., & Wang, T. (2014). Social media: A new vehicle for city marketing in China. *Cities*, *37*, 27–32.

Chapter 10

Tourism Operation in Kumbakarai Waterfalls: An Analytical Study

S. Kumaran[a,1], *B. Latha Lavanya*[b,2] *and T. Milton*[c,3]

[a]Alliance University, India
[b]MEASI Institute of Management, India
[c]Bharat Institute of Higher Education and Research, India

Abstract

Waterfalls are flow of water in the stream or river from mountains or hills to lower place. It is also called as cascades. Waterfall creation is significantly influenced by the process of erosion, which is the wearing away of the earth (Barton, 1974). It is usually located in the dense forest or bottom of the forest. The waterfalls are becoming important tourist destinations among tourists who are interested to visit forest through trekking (Goodall, 1990). Kumbakarai is the waterfalls located near Periyakulam town in the Vellagavi forest and it was developed by local businessman Chellan Iyer on par with Courtallam waterfalls in Tenkasi district. Kumbakarai waterfalls are being visited by thousands of tourists during the season and people from nearby districts visit frequently. This chapter mainly focuses on the tourist profile, mode of transport, facilities available in the waterfall area and suggestion for improving the destination to attract a greater number of tourists.

Keywords: Waterfall; tourist attractions; tourism elements; tourism facilities; transportation facilities

[1]https://orcid.org/0000-0003-1842-8999
[2]https://orcid.org/0000-0002-4185-9618
[3]https://orcid.org/0000-0003-4643-0981

Sustainable Tourism, Part B, 181–192

Published under exclusive licence by Emerald Publishing Limited
doi:10.1108/978-1-83608-326-920241010

1. Introduction

Kumbakarai waterfalls is situated at the foot of the Kodaikanal hill, near Periyakulam town, Theni District. It is located at an altitude of 1,300 feet. The number of drops of the waterfall is two and the height of the waterfall is 50 feet. The main water source for this waterfall is Kodaikanal hills and water flows along the rocks and finally reaches foothills at Kumbakarai. The exceeding water from Kodaikanal town reservoir will form the river and flows through a tourist attraction Fairy Falls and then it continues flows and reaches another tourist attraction Liril falls and finally reaches Kumbakarai waterfalls in the Vellagavi forest area. The waterfalls are located across the Pambar River. The water falls along the rocks in two stages. In the first stage the water collects in huge rock recesses that are named after wild animals, such as elephant, tiger, bison, wild boar and snakes. In the next stage, the water rolls off and falls to the base, forming a shallow pool at the bottom. From there, it strolls into a slow stream, before disappearing to the terrain again.

2. History

The water flows throughout the year and during the rainy season the intensity of flow will be heavy. Travelers visit this place for bathing throughout the year. Earlier this waterfall was surrounded by thick forest species and tourists' visits were very minimum. During 1842, Mr. Chellam Iyer, an affluent businessman and owner of Bhavani Krishna Vilas hotel, from Periakulam wanted to develop these waterfalls on par with Courtallam waterfalls. He sent a proposal to British Government and obtained permission from them in 1942 and constructed a dressing room for women, staircases and bathing platform at his own expenses. These facilities attracted many numbers of tourists to a famous and became famous tourist spot near Periakulam. He constructed a statue for Lord Muruga near waterfalls and tourists can see the stone inscription, which speaks about his contribution to these falls.

3. Importance

Kumbakarai waterfalls is a day trip tourism attraction and tourists enjoy by taking bath in the waterfalls between 8 a.m. and 4 p.m. A drive to Kumbakarai waterfalls along the lined trees, the sounds of gushing waters and singing birds, flying butterflies makes the tourists feel leisure and joy. It is nice place for photographic lovers. Whenever there is rain in Kodaikanal, one can observe the gushing water in the Vellagavi river, which leads to Kumbakarai waterfalls. People from 100 km radius visit this beautiful location especially during hot climate to battle the dreary summer. The sun light enters through the foliaged arches of tamarind, neem and jamun trees that line this route and throws shadows on the empty road. The colourful paintings on wildlife on the walls of the forest guest house and forest guard's office is another noticeable thing before getting to the waterfall. There is a beautiful National Medicinal Plants Board office cum park on spread over five hectares on the left side, which is a sample of Nature's bounty (Saranya & Mariswamy, 2016). The air in the park is energising and the silence is calming (Nimi, 2013). The birds chirp in the background like a distant chorus and the trembling butterflies speckle the green with colour. There is an observation tower across the stream that also serves as an ideal spot to view the waterspout. There is a stunning trekking route from here to Kodaikanal and one can reach top within 9–12 h depending upon the individual stamina.

4. Tourism Elements

The climate of the Kumbakarai waterfall is subtropical climate. It is located at an elevation of 1,300 feet. The temperature is ranging from 20°C to 35°C (Table 10.1).

Table 10.1. Kumbakarai Waterfalls: Tourism Elements.

S.N	Elements	Detail
1	Location	Located in the Foothills of the Kodaikanal hills at a 9 km distance from Periyakulam, Theni District Tamil Nadu state; 50 km from Kodai Road
2	Coordinates	10°10′48″N 77°31′50″E
3	Height	50 feet
4	Number of drops	Two
5	Watercourse	Pambar River
6	Climate	Subtropical high climate
7	Elevation	400 m/1,300 feet
8	Temperature	20–35°C
9	Precipitation	833 mm
10	Season	Summer: March–May, Monson: June–September and Winter: December–February. But water flows throughout the year. During rainy season water flow will be heavy.
11	Flora	Invasive alien species of plants such as wattle and pind.
12	Fauna	Elephants, Barking Deer, Indian Gaur, Sambar, Common Langur, Wild Dogs, Leopard and Sloth Bear
13	Attractions nearby	Deciduous Forest, Herbal Farm and Park.
14	Dress changing room	Available.
15	Separate bathing area for women	Not available.
16	Facilities at waterfalls	Dressing room, rest shed, camp shed, toilets are available.
17	Visiting hours	8.00 a.m.–4.00 p.m.

(*Continued*)

Table 10.1. *(Continued)*

S.N	Elements	Detail
18	Entry fee	For adult Rs.15/. For children Rs.10/
19	Bus facilities	Buses are available from Periyakulam. The timings are 6 a.m., 7 a.m., 1 p.m., 3.30 p.m. and 4.30 p.m. Auto and cab facilities are available.
20	Specialties	Medicinal plants and herbs/national medicinal plants park.
21	Lodging facilities	Hotels and lodges are available at Periyakulam.
22	Food facilities	Not available.
23	Approach road	Roads are good.
24	Parking	Parking places are available. For bike Rs.15, for car Rs.50 and for the bus Rs.100.
25	Guides	Professional guides are not available. But forest staff will help the tourists in waterfall area.
26	Hospitality of local people	Modest
27	Shop services	Sweet mango and hill honey is available in local market
28	Established	1942
29	Nearest city	Periyakulam 9 km, Dindigul 63 km and Madurai 90 km
30	Nearest airport	Madurai – 90 Km
31	Train facilities	Kodai Road – 50 km and Madurai – 90 km.
32	Contact address	District Forest Officer, Kodaikanal Division, Phone: 04542-240287. Email: kodaidfo@gmail.com

Source: Primary data.

The waterfalls area receives 833 mm rainfall annually. During the peak monsoon, the rain fall intensity is normally high and tourists are not allowed to take bath in the waterfall and rest of the time tourists are allowed to use it. The forest department staff will be available throughout the day and facilitate the tourists during their visits (Nanjundaiah, 2004).

Tourists are allowed to the falls through entrance ticket. Tourists are responsible in maintaining the forest ecology and keeping the waterfalls environment neat and clean (Tao & Wall, 2009). The wastes are put in the proper bin, which are kept around the waterfalls. The restrictions at Kumbakarai waterfalls are tourists should not enter the restricted forest area and no soap or shampoo is allowed in the waterfalls since it harms animals. Tourists are requested not to carry highly inflammable materials (Shanimol, 2004). Tourists (adult men) wear 3/4th or half pant while taking bath and washing of cloths is not allowed, and smoking and alcohol drinking are strictly restricted. The tourists are safeguarded by forest staff around the waterfalls throughout the day.

There is an exclusive dress changing room for ladies at Kumbakarai waterfalls and there is a toilet for tourists, and it will be as clean as expected by tourists. The dress changing room capacity is not adequate, because of the crowd. The entry fee for adult is Rs.15/ and that of for children is Rs.10/. Paid parking facilities are available at Waterfalls area and the charge for car parking is Rs.50, for Bus it is Rs.100/ and for bike it is Rs.15/. To reach this falls, buses, autos and cab facilities are available from nearby town Periyakulam. Food facilities are not available at waterfalls area. The lodging facilities and hotels are available for the tourists at Periyakulam. There are no professional guides available, but forest department staff facilitates the tourists in the waterfall area. Tourists purchase sweet mango and hill honey at local market in Kumbakarai as well as in Periayakulam. The nearest airport is available at Madurai, which is at 90 km distance from the waterfall and train facilities are available at Kodai Road and Madurai.

5. Tourist Attractions

The tourist attractions nearby Kumbakarai waterfalls are Vaigai Dam, a water reservoir providing irrigation to the command area in the downstream, Suruli waterfalls, which about 66 km from Kumbakarai, the Kodaikanal hills station, which is located 75 km from Kumbakarai and Sathurangapara viewpoint, a breath-taking viewpoint located at the border of Theni and Idukki district of Kerala (Table 10.2).

Table 10.2. Tourist Attractions Nearby.

S.N	Name of the Attractions	Attraction Detail
1	Vaigai Dam	
2	Suruli Waterfalls	It is the beautiful two stage cascading waterfalls drop from a height of 150 feet, located 66 km from Kumbakarai. The water from the falls has medicinal value and cures various pains. Near the falls there are five caves which represents the Indian Rock Architecture of 11th century.
3	Kodaikanal	It is a beautiful hill station, located 75 km from Kumbakarai. It has tourist attractions like Lake, Green valley view, waterfalls and Grassy hills.
4	Chathurangapara viewpoint	Mesmerising and breath-taking viewpoint located at the border of Theni and Idukki district of Kerala. It is a gorgeous hiking spot. There are hundreds of windmills located on the viewpoint, because wind velocity is more enough to generate electricity through windmill.

Source: https://theni.nic.in/tourist-place/kumbakaraifalls/

மூலிகை வனம், கும்பக்கரை அருவி.

Source: https://en.wikipedia.org/wiki/Kumbakkarai_Falls; https://theni.nic.in/tourist-place/kumbakaraifalls/

The *main objectives* of this chapter are to assess the tourists visiting Kumbakarai waterfalls and find out the facilities and service available in the waterfalls area. It also aims to assess the cleanliness in the tourism area. Based on the research findings, it provides recommendations for the improvement of the tourism place.

Scope of the study is restricted to tourists visiting the Kumbakarai waterfalls for leisure and entertainment. The study covers the tourism infrastructure, cost of facilities, hygiene and the happiness of the tourists. It seeks to provide useful suggestions for the overall improvement of the tourism activities in the Kumbakarai area.

6. Research Methodology

The study is descriptive in nature and is based on data that the researchers collected from primary sources by way of a sample survey. Information about waterfalls is also gathered from secondary sources including research articles, journals, novels and magazines; 100 people made up the sample. In his research, open, non-probability convenience sampling procedures were employed. Tourists from various regions have been selected for this study. A questionnaire has specifically been designed to gather information from the tourists.

7. Accompany Details

The researcher makes an effort to look at those that accompany tourists. Twenty-two percent of travellers are accompanied by their partners, and 66% are family members; 12% of groups are mixed (Table 10.3).

Table 10.3. Accompanied Persons.

Accompany Details	No. of Tourists	%
Spouse	22	22
Family affiliates	66	66
Mix up group	12	12
Total	100	100

Source: Primary data.

The above table exposes that majority of the tourists are accompanied by Family members.

7.1 Visit Purpose

The researcher is aware that the reason for tourists' visits is a crucial element in the growth of the tourism industry. Out of 100 tourists surveyed, 86% are travelling just for pleasure, 6% are travelling for both work and pleasure and the remaining 20% are travelling to see friends and family (Table 10.4).

According to the report, 86% of visitors are solely there for recreational purposes.

7.2 Count of Visits

The study gathered information on the number of tourists who visited Kumbakarai Falls. Approximately 76% of visitors to Kumbakarai Falls are first timers, 20% for the second time and the remaining 4% has visited falls more than twice (Table 10.5).

7.3 Transport

The study investigates the method of transportation used by visitors to Kumbakarai Falls. Of the 100 visitors surveyed, 34% drove their own vehicle to the falls, 52% took a bus and 14% used a rented car (Table 10.6).

Table 10.4. Visitation Goals.

Particulars	Tourists in Number	Percentage
Pleasure	86	86
Work and pleasure	06	06
Visit to friends and relatives	8	8
Total	100	100

Source: Primary Data.

Table 10.5. Visit to Waterfall.

Number of Times	Number of Tourists	Percentage
First timers	76	76
Two	20	20
More than two	04	04
Total	100	100

Source: Primary Data.

Table 10.6. Mode of Transport.

Number of Times	Number of Tourists	Percentage
Possess a vehicle	34	34
Bus	52	52
Vehicles rented	14	14
Total	100	100

Source: Data from Primary sources.

According to the research, the majority of tourists choose the bus as their form of transportation to get to Kumbakarai Falls. The development of tourism is dependent on the transportation options offered to visitors at their location (Su et al., 2015).

The tourism facilities in Kumbakarai falls include Transport, parking, security, bathing facilities at waterfalls, dress changing room, food courts, toilet facilities and local people hospitality. 'parking' at waterfalls scored highest points (321) and ranks first followed by 'bathing facilities' (244 points), road and transport (227 points), 'dress changing room' (218 points), 'toilets' (205 points), 'Local people hospitality' (198 points) and 'food courts' and 'security' (195 points). From the table it is understood that the security arrangements are not to the expected level and the food courts are hygienic and clean (Table 10.7).

Findings.

- The majority (66%) of visitors are accompanied by family members, and the majority (86%) of visitors are there only for leisure.
- The majority (76%) of tourists are first-time visitors to the falls.
- Fifty-two percent of travellers used bus transportation to get to their destination, according to the report.
- The visitors gave Kumbakarai Falls the following ratings: 'Parking' at the waterfalls received the highest marks (321), followed by 'bathing facilities' (244), 'road and transport' (227), 'dress changing room' (218 points), 'toilets' (205 points), 'Local people hospitality' (198 points), 'food courts' and 'security' (195 points).

Table 10.7. Opinion About the Tourism Facilities in Kumbakarai Falls.

	Opinion											
	Very Good (5)		Good (4)		Moderate (3)		Poor (2)		Very Poor (1)			
Facilities	**Number**	**Score**	**Number**	**Score**	**Number**	**Score**	**Number**	**Score**	**Number**	**Score**	**Total**	**Rank**
Security	0	0	0	0	25	75	45	90	30	30	195	7
Bathing facility	0	0	0	0	55	165	34	68	11	11	244	2
Food court	0	0	0	0	25	75	45	90	30	30	195	7
Road and transport	0	0	0	0	51	153	25	50	24	24	227	3
Dress changing room	0	0	0	0	32	96	54	108	14	14	218	4
Toilets	0	0	0	0	23	69	59	118	18	18	205	5
Local people hospitality	0	0	0	0	19	57	60	120	21	21	198	6
Parking	10	50	18	72	55	165	17	34	0	0	321	1

Source: Primary data.

- The dress changing room for women is not maintained properly.
- Cleanliness is not maintained in the water fall area.
- Safe drinking water for the visitor is not arranged.
- The restrooms are not properly maintained.
- The waste bottles are thrown near the parking area.
- Food waste is not properly put in the dust bin.

Suggestions.

- There should be an exclusive website for Kumbakarai waterfalls which depicts information like location, season, road and transport facilities, information about waterfalls and their photos.
- An exclusive information centre should be created at the bus stand and railway station to facilitate the tourist to have a successful tourism plan.
- An advertisement signboard should be installed in important cities in Tamil Nadu to draw more visitors.
- Since the majority of the visitors are local people, the bus facilities and road facilities approaching the waterfalls should be improved.
- Separate bathing facilities for women need to be arranged in the waterfall area.
- More dress changing rooms must be provided near the waterfall for women.
- There is an arrangement for safe drinking water and a roofed area for taking lunch for the visitors.
- The Forest department and tourism department should collectively plan to develop the waterfall area and maintain it properly.
- Since the present restrooms are not adequate, a greater number of restrooms should be constructed.
- The cleanliness and hygienic condition of the food courts have to be improved.
- A guest house may be constructed by the forest department near the waterfalls to attract a greater number of tourists.
- Tamil Nadu Tourism Development Corporation (TTDC) should take steps to open a hotel in Kumbakarai to attract more number of tourists. The proposed hotel should have conference hall, dormitory and restaurant (The Hindu, 2018).

9. Conclusion

Waterfall tourism is gaining momentum among tourists (Singh, 2010). In Kumbakarai falls throughout the year the water is flowing and visitors are mostly from local and surrounding areas. The tourist inflow can be increased by familiarising the waterfalls in Vellagavi forest through a proper advertising programme. The research study on Kumbakarai falls has found certain issues pertaining to facilities and services. A few suggestions are provided based on the findings and by implementing the suggestions effectively, the tourist inflow can be increased and the socio-economic condition of the area will be improved (Karr, 1999).

References

Barton, R. M. (1974). *Waterfalls of the world. A pictorial survey*. D. Bradford Barton Ltd.

Goodall, B. (1990). The dynamics of tourism place marketing. In G. Ashworth & B. Goodall (Eds.), *Marketing tourism places* (pp. 259–279). Routledge.

Government of Tamil Nadu. Kutladampatti Waterfalls, Accessed on 11 December 2018.

Karr, J. R. (1999). Defining and measuring river health. *Freshwater Biology 41*, 221–234.

Life of Swami Ramanagiri. David's Godman blog.

Nanjundaiah, C. (2004). *Forests, environment and local community – A case study of Kodagu (Coorg) District*.

Nimi, D. R. (2013). *Marketing strategies adopted by Tourism Industry in South India – A Comparative study between Kerala and Tamil Nadu in her study*. Published thesis.

Saranya, T., & Mariswamy, H. K. (2016). The role of health tourism on tourism development in Kerala State: An evaluation. *International Journal of Management and Social Sciences*, *4*(2), 494–501. https://research.advances.org/index.php/RAJMSS/article/view File/453/472

Shanimol, S. P. (2004). *The educational potentialities of tourism development in Kerela and Tamil Nadu*. Manonmaniam Sundarnar University. Unpublished dissertation.

Singh, J. (2010). *Ecotourism*. I.K. International Publishing House Private Limited.

Su, M., Wall, G., & Xu, K. (2015). Heritage tourism and livelihood sustainability of a resettled rural community: Mount Sanqingshan World Heritage Site, China. *Journal of Sustainable Tourism*, *24*(5), 735–757.

Tao, T. C. H., & Wall, G. (2009). Tourism as a sustainable livelihood strategy. *Tourism Management*, *30*(1), 90–98.

The Hindu. (2003, January 20). *Tourism, Madurai's mainstay*. Archived *from the original on 13 November 2009*. Accessed on August 4, 2009.

The Hindu. (2018, April 3). *Varkala emerges prime tourist destination* [News]. https://www.thehindu.com/news/national/kerala/varkala-emerges-primetourist-destination/article23428045.ece

Vijay Anand, A. S. Impact of tourism on environment: A study in Ooty town. Mononmaiam Sundarnar University. Unpublished dissertation.

Chapter 11

Solar Brilliance: Illuminating the Modhera Sun Temple With Sustainable Energy

Kingshuk Biswas

Gujarat Power Corporation Limited, India

Abstract

The Modhera Sun Temple, an architectural masterpiece dating back to the Chaulukya dynasty's reign in 1026-27 CE, graces Gujarat's Mehsana district. This historical gem holds profound cultural significance, and the 'Solarisation of Modhera Sun Temple and Town' project draws inspiration from this iconic monument. Its core vision is to seamlessly blend the temple's historical importance with modern sustainability by harnessing solar energy to power Modhera town. At the heart of this transformative endeavour lies the establishment of a 6 MW Grid-Connected Ground Mounted Solar PV Power Plant in Sujjanpura. This is complemented by the installation of 1383 1KW Rooftop Solar Systems on residential buildings in Modhera village. To ensure a continuous power supply, a 15 MWh Battery Energy Storage System (BESS) in Sujjanpura addresses solar power intermittency. The project also incorporates smart energy metres, a SCADA-based Smart Energy Centre at the Modhera Sun Temple, a 50 KW Solar Parking Infrastructure with Battery Storage and Electric Charging Stations, and sensor-based smart street lights. In October 2022, the Honourable Prime Minister inaugurated the Solar Powered 3-D Projection Mapping Show and Heritage Lighting at the Temple. Serving as a model for rural electrification through solar energy, it offers valuable insights for tackling renewable energy challenges on a broader scale. Noteworthy achievements include Modhera becoming the first Indian village to enjoy uninterrupted 24×7 solar energy access, a testament to the seamless fusion of ancient heritage and modern sustainability.

Keywords: Modhera Sun Temple; solarisation; renewable energy; sustainability; cultural heritage; rural electrification; smart energy solutions

Sustainable Tourism, Part B, 193–207

doi:10.1108/978-1-83608-326-920241011

1. Introduction

Within the context of Indian culture, the sun holds a position of profound reverence, owing to its role in sustaining life through the provision of warmth and energy to all living organisms on Earth. This ancient tradition of venerating natural elements and recognising their interconnectedness with human existence underscored the significance of solar energy as a vital force within the life cycle. India, endowed with abundant sunlight, continually reaps the manifold benefits bestowed by this celestial source. In the pursuit of diminishing the nation's reliance on conventional energy sources, many of which bear environmental and economic burdens, a critical question emerged: Could the iconic Modhera Sun Temple be illuminated using solar power? It was at this juncture that the notion emerged that there could be no more fitting expression of devotion than to adorn the resplendent precincts of the Sun Temple and the surrounding village with the radiance of solar energy. Solar power stands out as an exceptionally clean and boundless energy source, uniquely poised to address India's extensive energy demands while concurrently providing a safeguard for both the nation and the environment.

The Modhera Sun Temple stands as a remarkable architectural masterpiece, serving as an enduring symbol of India's rich cultural heritage. Situated in the picturesque Mehsana district of Gujarat, this ancient marvel represents a blend of history and spirituality. Constructed during the reign of Bhima I of the Chaulukya dynasty in 1026-27 CE, the Modhera Sun Temple has been a source of wonder and inspiration for generations.

At the entrance of the historical complex is the magnificent pond (kund) known as the Ramakund, built in a rectangular shape and containing 108 shrines dedicated to various gods and demigods. Three main shrines are positioned on the three sides of the kund, dedicated to Ganesh, Vishnu and an image of Lord Shiva dancing the 'tandav'. Visitors can then ascend the steps to the 'SabhaMandap', the assembly hall, adorned with sculpted renderings of 12 'Adityas', representing the sun according to the 12 months.

After witnessing the 'Sabha Mandap', a visit to the Garbhagruha will undoubtedly be a deeply spiritual experience for the visitor. Although the statue of the sun god no longer exists, the temple continues to evoke the legacy of the Suryavanshi Solanki dynasty. Remarkably, on the day of the Equinox, the temple resonates with the echoes of prayers, the fragrance of incense and the gentle tinkling of bells. During this transcendent moment, the sharp, linear rays of the sun converge to illuminate the inner sanctum, symbolising the eternal connection between life and light.

The nearest railway station to the Modhera Sun Temple is the Mehsana Junction, which is well-connected to major cities across India, serving as a vital transportation hub for those embarking on a journey to explore this site. Additionally, for those arriving by air, the Sardar Vallabhbhai Patel International Airport in Ahmedabad, located approximately 102 km away from Modhera, offers convenient access. This international airport connects travellers to both domestic and international destinations, making it an accessible gateway for tourists eager to visit the Modhera Sun Temple.

2. Literature Review

2.1 Renewable Energy in Tourism Sector

According to a press release by Petroleum Minister Hardeep S. Puri (2023), India's energy demand is growing very quickly. It is expected to make up 25% of the world's energy demand increase over the next 20 years. This shows the need for sustainable energy solutions in the tourism sector, like solar power. India's energy use is nearly three times the global average. Solar energy is a practical solution for the tourism sector, especially in rural areas. It provides a reliable and sustainable energy supply while also reducing carbon emissions. The world is moving towards using more renewable energy. This supports the use of solar energy in tourism, which aligns with global sustainability goals and reduces reliance on fossil fuels.

According to Ben Jebli et al. (2019), sustainable tourism, as defined by the United Nations Environment Programme (2011), encompasses tourism practices that aim for long-term viability within economic, social, cultural and environmental frameworks. It is not a distinct form of tourism but rather an aspiration for all types of tourism to prioritise sustainability. Ecotourism, a subset of the industry, specifically centres on environmental sustainability, striving to safeguard natural resources through enhanced energy efficiency and increased use of renewable energy sources. Investments in public goods such as waste management and renewable energy infrastructure can significantly reduce the expenses for private sector involvement in environmentally conscious tourism. Governments can facilitate green tourism initiatives by offering tax breaks, subsidies or time-limited incentives for adopting waste reduction, energy efficiency or renewable energy technologies. It is essential to accurately assess the actual societal cost of resource use, energy consumption and waste production. The United Nations Environment Programme (2011) recommends integrating sustainability as a focal point in tourism promotion and marketing strategies. Public-private partnerships are encouraged to distribute the costs and risks associated with substantial investments in green tourism. Governments can further support these initiatives by lowering administrative fees and providing assistance in technical, marketing or business domains. The intricate relationship between tourist numbers and renewable energy consumption suggests that policies boosting the tourism sector might serve as a catalyst for expanding the use of renewable energy sources. Conversely, promoting renewable energy usage can attract more tourists to these regions. Additionally, Foreign Direct Investment (FDI) offers economic and environmental benefits to these nations. To attract more FDI, countries should enhance infrastructure, workforce qualifications and administrative procedures.

According to Kuldasheva et al. (2023, p. 150), countries heavily reliant on tourism often contribute to environmental degradation by depending on non-renewable energy sources. Achieving sustainable economic development necessitates a rapid transition to renewable energy technologies in these regions. Existing research underscores the dual advantages of renewable energy, both in environmental conservation and fostering economic growth. Policymakers are advised to simplify administrative processes associated with the adoption of renewable energies by households and companies. The development of public–private partnerships (PPPs) is advocated to boost investments in renewable energy sources, enabling more active involvement of private investors in renewable energy initiatives. Addressing the information gap regarding the benefits of clean energy is crucial for enhancing the feasibility of such projects. Future studies should explore additional variables that could play a significant role in predicting carbon dioxide (CO_2) emissions in tourism-dependent markets. Variables like human capital and institutional quality warrant thorough examination to offer a comprehensive understanding of the multifaceted aspects involved.

According to Rudiany et al. (2023), the coupling of renewable energy and tourism in Southeast Asia offers prospects for local economic advancement, yet it is fraught with challenges such as maintaining renewable technology, issues of overexploitation and misdirected energy utilization. This collaboration does not invariably guarantee sustainable tourism outcomes. The introduction of 'energy culture' emerges as crucial, delving into how societies understand and utilise energy sources, encompassing cognitive norms, material culture and energy practices. Southeast Asia confronts environmental stress from energy over-exploitation, necessitating a reassessment. Energy culture provides a means to align human activities with nature. Cognitive norms encourage reevaluation of sacred sites before implementing renewable energy projects, aligning with societal veneration for the environment. Material culture constrains resource use based on availability and conditions, while energy practices adhere to traditional values, mitigating overexploitation. For sustainable tourism, comprehensive societal involvement is essential. Policies integrating renewable energy must consider socio-cultural values, emphasising respect for nature and local energy sources. Respect, harmonisation and inclusive participation should steer energy alignment to ensure environmentally and socially acceptable, economically viable tourism in Southeast Asia. Aligning cognitive norms, material culture, and energy practices within the energy culture framework fosters a unified approach.

VR (2023) in his blog has discussed solar energy, derived from solar radiation and harnessed through devices like solar panels and photovoltaic cells, holds significant promise in the quest to combat climate change and reduce greenhouse gas emissions. Its renewable and pure nature aligns seamlessly with the principles of sustainable tourism, which emphasizes the consideration of fiscal, social and environmental impacts in travel. Sustainable tourism practices involve responsible actions to minimize waste, pollution and environmental degradation while supporting local communities. The integration of solar energy into the tourism sector, particularly in accommodations like hotels and resorts, can effectively contribute to environmental sustainability. Solar cells strategically placed on rooftops, parking areas, and water surfaces can generate energy and provide hot water for various amenities. Given that the tourism industry contributes approximately 8% of global greenhouse gas emissions, addressing its environmental impact is critical. Primary sources of carbon emissions in this sector include fossil fuel consumption for transportation, heating and cooling of accommodations, and tourist-related activities. The adoption of solar energy emerges as a potent solution to reduce the carbon footprint by providing cleaner alternatives for electricity generation, water heating and space conditioning.

While the initial investment in solar technology may be substantial, the potential for significant savings over time makes it an attractive and sustainable option for these businesses. Examining the impact of solar energy on local communities within tourism destinations reveals several positive outcomes. Firstly, solar energy contributes to cost savings for local communities, freeing up resources for other initiatives. Secondly, it promotes sustainability in tourism, appealing to environmentally conscious tourists. Thirdly, solar energy initiatives create opportunities for local economic development, offering employment and education prospects. Fourthly, it enhances energy independence, particularly in remote areas with limited grid access. Fifthly, solar energy increases adaptability to natural disasters and infrastructure disturbances, ensuring the continuous operation of essential services. Sixthly, it contributes to education opportunities, fostering public awareness of renewable energy sources. Finally, solar energy projects encourage community involvement, fostering collaboration and strengthening interpersonal ties within the community.

2.2 Solar Energy and Heritage Conservation

According to Cabeza et al. (2018, p. 103), integrating solar energy into historical buildings poses significant challenges, primarily due to space constraints. Another critical concern is the necessity to preserve the architectural integrity of the building. This difficulty is further emphasised by the need for interventions to be visually unobtrusive. One promising avenue for integration involves the utilization of new photovoltaic tiles, which are comparatively less visually intrusive than alternative retrofit systems. By adopting such technologies, it becomes possible to mitigate the carbon emissions associated with building operations.

According to Christiernsson et al. (2021), the installation of solar cells, especially on rooftops, presents a challenge as it can significantly change a building's visual aesthetics. When these installations obscure the original roofing material, there is a risk of compromising the cultural and artistic significance of the structure. In specific local planning zones, acquiring a building permit becomes necessary when solar cell installations significantly alter buildings with distinct cultural, artistic, or aesthetic values. This requirement extends to structures situated in areas designated for their cultural significance. Notably, installations on high, flat roofs are often viewed as less disruptive and are more likely to be considered permissible, as they tend to have a lesser impact on a building's overall appearance.

According to Polo López and Frontini (2013), to mitigate the uncontrolled proliferation of solar technologies and optimise the allocation of economic and spatial resources, it is imperative to strike a careful equilibrium between technical functionality and aesthetic considerations. The successful integration of Building-Integrated Photovoltaics (BiPV) or solar thermal systems demands a heightened emphasis on architectural design quality, as evidenced by prior experiences (Times of India, 2016). In the current market, numerous innovative products tailored for integration into sensitive built environments are available. The incorporation of renewable technologies into urban landscapes or existing structures holds the promise of enhancing both architectural and technical aspects, thereby contributing to economic and environmental sustainability.

However, challenges arise from scenic parameters that presently impede the adoption of these technologies in towns and historical downtown areas from harnessing the associated benefits. Striking a delicate balance between these considerations is crucial to achieving successful integration and realizing the full potential of renewable technologies.

3. Energy Consumption in India

As per the Central Electricity Authority (CEA), the per capita consumption of energy in India for the year 2022 stands at 1255 Kwh. For the last 5 years, on an average the CAGR has remained at 2.35%, as shown in Fig. 11.1. India's energy demand will continue to provide fuel for future economic growth and is bound to grow exponentially in the coming years. India is likely to account for 25% of global energy demand growth over the next two decades.

At present, the country is the world's third largest consumer of oil, third largest LPG consumer, fourth largest LNG importer, fourth largest refiner and fourth largest automobile market. Also as per the latest report of Central Electricity Authority (CEA) for August 2023, India has installed power generation capacity of 424 GW, which includes 206 GW coal based, 47 GW of large hydro and about 132 GW renewables (solar, wind energy, etc.).

4. Gujarat Renewable Energy Scenario

According to the Ministry of New and Renewable Energy, Government of India (2023) as of 31 August 2023, Gujarat stands as a significant player in India's renewable energy landscape with a total installed capacity of 22,435 MW. This

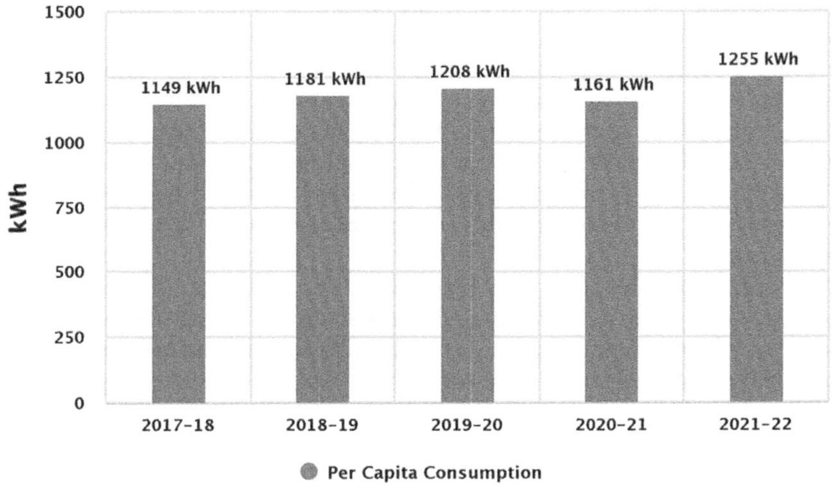

Fig. 11.1. Per Capita Energy Consumption in India. *Source:* Central Electricity Authority of India. (2023, November 29). Dashboard.

impressive figure comprises 10,395 MW of solar, 11,060 MW of wind, 110 MW of bio-power, 90 MW of small hydro and 770 MW of large hydro capacity. Remarkably, approximately 46.24% of Gujarat's total installed capacity is sourced from renewable energy. The state secures its position as the leader in wind energy with an impressive 11.1 GW, while also claiming the second position in solar energy with 10.4 GW. Gujarat ranks second in total renewable energy installed capacity nationwide, standing strong at 22.5 GW, with wind and solar sectors demonstrating commendable Compound Annual Growth Rates (CAGRs) of 14.0% and 34.0%, respectively. Furthermore, Gujarat excels in rooftop solar installations, boasting a substantial 2.9 GW capacity, benefiting 4.4 lakh beneficiaries. These statistics underscore Gujarat's pivotal role in advancing renewable energy solutions.

5. The Project

In June 2019, the Government of India and the Government of Gujarat have collaboratively initiated the 'Solarisation of Modhera Sun Temple and Town' project, designed to provide continuous solar energy (24×7) to Modhera town. This is achieved through a Solar Power project integrated with a Battery Energy Storage System situated in Village Sujjanpura, District Mehsana and approximately 6 km from the renowned Sun Temple in Modhera. The project spans 12 hectares of land and encompasses the following key components:

(a) A 6 MW Grid-Connected Ground-Mounted Solar Photovoltaic (PV) Power Plant located in Sujjanpura (Fig. 11.2). (b) Installation of more than 1,300 × 1 kW Rooftop Solar Systems on residential buildings in Modhera, Samlanpura, and Sujjanpura villages. (c) Implementation of 316 kW Rooftop Solar PV Systems on various government buildings, including schools,

Fig. 11.2. 6 MW Solar Power Plant at Sujjanpura Village.

government offices, community centres and the bus stand in Modhera, Samlanpura and Sujjanpura villages. (d) Deployment of a 15 MWh, 6 MW Battery Energy Storage System (BESS) in Sujjanpura. (e) Integration of Smart Energy Components, including:

- Installation of over 1,700 Smart Energy Metres at the consumer level in Modhera.
- Establishment of a SCADA-based Smart Energy Centre, serving as an Interpretation Centre cum Museum, at the Modhera Sun Temple.
- Development of a 50 kW Solar Parking Infrastructure featuring a 150 kWh Battery Storage and Electric Charging Stations at the Modhera Sun Temple (Fig. 11.3).
- Implementation of Sensor-based Smart Street Lights in the vicinity of the Modhera Sun Temple.

The project was dedicated to the nation on 9th October 2022 by the Hon'ble Prime Minister of India. The Modhera village is now immersed with never-before felt energy of the Sun with conversion of solar energy into power to energise the whole village. The entire Modhera village has now been relieved of the burden on conventional coal based electricity and thereby championing the usage of solar energy.

Fig. 11.3. Modhera Sun Temple Parking Area.

6. Stakeholders Involved

The Solarisation of Modhera Sun Temple and Town project is truly collaborative. The role of every stakeholder was paramount in the successful implementation of the project.

Ministry of New & Renewable Energy: Apart from generous funding support, the ministry offers crucial technical guidance and knowledge-sharing throughout the project.

Gujarat Power Corporation Limited: Serving as the project's owner and implementer, GPCL is responsible for the timely establishment of all project components.

Archaeological Survey of India: This organisation plays a pivotal role by granting timely permissions and providing guidelines for the installation of the 3D project mapping show.

Tourism Corporation of Gujarat Limited: This body funds the 3D project mapping show and oversees the implementation of various tourist-centric facilities in and around the Temple complex.

Mehsana District Administration: Given that the project is situated in the district, efficiently managed all clearances, licences and related project matters.

7. Energy Needs of the Solar-Powered Sun Temple

Before the installation of the 3D projection mapping show at the Sun Temple, the temple's energy consumption stood at approximately 700 kWh, primarily attributed to the operation of pathway lights, high mast lighting, museum facilities, restroom blocks and office spaces. Subsequently, the State Tourism Department introduced the 3D projection mapping show as part of their revitalisation efforts for the Sun Temple. This endeavour involved the deployment of projectors equipped with weather-resistant outdoor architectural lighting, aimed at illuminating the temple's façade and creating a reflective spectacle on the adjacent body of water. The energy consumption associated with the 3D mapping show was estimated to be around 50 kWh. Before the implementation of solar panels, the monthly electricity bill for the Sun Temple premises amounted to approximately Rs. 5,000 (net metring with solar tariff @ Rs.1.75). The Solarisation of Modhera Sun Temple and Town project, undertaken by the Gujarat Power Corporation Limited, also encompassed the installation of solar modules for the Sun Temple. Specifically, the temple grounds were equipped with a 25 kW rooftop solar array, consisting of 75 numbers with 335 W solar modules (Fig. 11.4). Following the completion of the solarisation project, the energy expenses for the Solar Modhera Temple, inclusive of the 3D projection show, have not only been reduced to zero but have also led to surplus solar power being fed into the state's electricity grid. As of the present date, the 25 kW solar system mentioned above has effectively contributed to a reduction of approximately 34,262.7 kg of CO_2 emissions.

Fig. 11.4. 25 KW Rooftop Solar Array on Modhera Sun Temple Complex Premises.

8. Impact of Energy Self-Efficiency in Modhera Village

The achievements of the Suryagram Project in Modhera are truly remarkable, making it the first village in India to achieve the status of a net renewable energy generator. The installation of rooftop photovoltaic (RTPV) systems by the Gujarat government, as part of the Suryagram Project, has significantly reduced power expenses for the majority of households in the village.

With approximately 1,700 houses in Modhera, 1,350 of them now boast RTPV systems. These houses are interconnected with the central electricity grid, resulting in a remarkable reduction of power expenses ranging from 40% to 100% for most households. Impressively, for the past 6 months, the village has not needed to import any power from the grid, and instead, it has been contributing green power back to the grid. Modhera generates about 6,332 kWh of power daily. The village's daily consumption stands at approximately 6,000 kWh, leaving a surplus that is supplied to the grid. Furthermore, solar panels have been installed on government buildings, including the iconic Sun Temple, emphasising a commitment to sustainable practices.

The Sun Temple, now powered by solar energy, also hosts an electric charging facility for motor vehicles. This free service is frequently utilised by tourists. Daytime power is sourced from solar panels, while electricity after sunset is supplied through the Battery Energy Storage System, located in Sujjanpura village, about 6 km from Modhera. The positive impact of the Suryagram Project is particularly significant for the residents of Modhera, most of whom are small farmers, agricultural laborers or livestock herders. The average household income, ranging from Rs. 8,000 to 10,000, suggests that the project has the potential to substantially enhance the quality of life in the village.

Additionally, the presence of educated households and the employment of many youth in nearby automobile factories highlight the diverse demographics and potential for economic development in Modhera. The village's successful transition to a net renewable energy generator sets a promising example for other communities, both in India and globally, looking to adopt sustainable and environmentally friendly practices.

9. Project Beneficiaries

The primary beneficiaries of the Solarisation of Modhera Sun Temple and Town project are unequivocally the residents of Modhera village, which has a population of 6,500. Following the project's implementation, the entire electricity requirements of the village are effectively fulfiled through solar energy. This provision extends beyond daylight hours, as a state-of-the-art energy storage system ensures uninterrupted access to solar power during the night. Consequently, Modhera village proudly holds the distinction of being India's inaugural 24×7 solar-powered village.

In addition to the centralised solar plant, more than 1,300 households in Modhera and neighbouring areas have been equipped with 1 KW rooftop solar panels (Fig. 11.5). Consequently, these households have become autonomous solar energy generators, catering to their own energy needs. Furthermore, leveraging the concept of net metring, any surplus energy generated by these

Fig. 11.5. 1 KW Solar Rooftop Modhera Village Beneficiary.

rooftop solar installations is seamlessly integrated into the state electricity grid. This ingenious mechanism not only translates into zero electricity bills for many households but also results in government credits to their bank accounts for the surplus solar power they contribute to the grid. The Solarisation of Modhera Sun Temple and Town project exemplifies a quintessential model of sustainable tourism. This iconic tourist destination has transitioned to being entirely powered by solar energy, marking a significant shift towards eco-friendliness and sustainability. Notably, the temple premises now feature state-of-the-art electric vehicle (EV) charging stations, catering to tourists of all vehicle sizes, thus actively promoting clean and green energy practices (Fig. 11.6).

The surge in tourist footfall at Modhera in recent years tells a compelling story:

- In 2020–21, there were 0.056 lakhs of tourists.
- By 2021–22, this figure had soared to 2.20 lakhs.
- In 2022–23, it further escalated to 3.78 lakhs.

These statistics vividly demonstrate that the introduction of the 3D mapping show at the temple has drawn significant crowds, particularly in the evening, as tourists flock to witness the captivating sound and light presentation (Fig. 11.7). The State Government plays a proactive role in encouraging school visits to this spectacular show. The impact extends beyond tourism. The Modhera village, situated adjacent to the temple, has achieved self-sufficiency in its energy needs through solar power adoption. This transition has translated into substantial

Fig. 11.6. Solar Powered EV Charging Station in Modhera Sun Temple Parking Area.

Fig. 11.7. 3D Mapping Sound and Light Show at Modhera Sun Temple.

reductions in monthly electricity bills for village residents, thereby empowering them financially.

In essence, the project effectively addresses and enhances the three core pillars of sustainable tourism: environmental considerations, economic benefits for the community and the preservation of the destination's cultural identity.

10. Achievements

This project represents a pioneering and successful initiative, serving as a noteworthy demonstration of rural self-sufficiency through the adoption of sustainable green energy, particularly Solar Power. Several significant milestones have been achieved:

- Modhera village has earned the distinction of being the first village in India to attain uninterrupted access to solar energy on a 24/7 basis.
- The Sun Temple in Modhera now exclusively relies on solar energy to power its 3D projection mapping show.
- Modhera village has achieved a remarkable status as India's inaugural village to transition into a net renewable energy generator.
- It stands out as the first contemporary village in India equipped with a state-of-the-art electric vehicle charging station powered by solar energy.

- The village of Modhera is home to India's pioneering grid-connected MWh Scale Battery Energy Storage System.
- Over 1,300 households in Modhera have embraced sustainability by installing rooftop solar systems, resulting in substantial electricity bill savings ranging from 60% to 100%.

11. Conclusion

The solarisation of Modhera Sun Temple and Town project stands out as a unique endeavour, not only within India but on a global scale. In the realm of remarkable tourism initiatives, it is often the tourism authorities or closely related bodies that take the lead in project commissioning. Typically, these projects are conceived within the purview of tourism authorities, which operate with separate budget allocations, sometimes resulting in isolated project developments. However, the dynamics are changing, particularly in a post-COVID-19 context, with a growing emphasis on sustainable tourism practices. In this evolving landscape, tourism projects are increasingly recognized as inter-sectorial endeavours that encompass critical dimensions such as climate change mitigation, social inclusivity and innovative design. Notably, community engagement has emerged as a pivotal aspect of sustainable tourism practices. Consequently, there is a growing imperative for cross-sectoral collaboration, drawing expertise from domains such as renewable energy, rural development, healthcare and education, among others. The renewable energy sector, in particular, has assumed a paramount role within the framework of tourism projects, given the heightened emphasis on carbon emission reduction and sustainability.

This project serves as a compelling example of the essential integration of renewable energy within the tourism sector. Encouragingly, similar initiatives are underway across India, signifying a broader trend in the tourism industry:

- The famed Sanchi Stupas in Madhya Pradesh, which is a very attractive tourist destination, is now known as 'India's first solar city'.
- The equally famous Konark Sun Temple and Town in Odisha will soon be fully powered by solar energy.
- Archaeological Survey of India is actively planning to set up solar power units across all its protected monuments in India. Depending upon the requirements at the site, it is proposed to set up solar power plants between 5 MW and 25 MW.

These collective efforts reflect India's proactive strides towards embracing sustainable tourism practices by incorporating renewable energy into its tourism projects.

References

Ben Jebli, M., Ben Youssef, S., & Apergis, N. (2019). The dynamic linkage between renewable energy, tourism, CO_2 emissions, economic growth, foreign direct investment, and trade. *Latin American Economic Review*, *28*, 2.

Cabeza, L. F., de Gracia, A., & Pisello, A. L. (2018, July 25). Integration of renewable technologies in historical and heritage buildings: A review. *Energy and Buildings*, *177*, 96–111.

Central Electricity Authority of India. (2023, November 29). *Dashboard.*

Christiernsson, A., Geijer, M., & Malafry, M. (2021). Energy efficiency, cultural heritage values and the law – Conflicts and potential solutions. *IOP Conference Series: Earth and Environmental Science*, *863*(1), 012026.

Kuldasheva, Z., Ahmad, M., Salahodjaev, R., & Fahlevi, M. (2023). Do tourism and renewable energy influence CO_2 emissions in tourism-dependent countries? *International Journal of Energy Economics and Policy*, *13*(6), 146–152.

Ministry of New and Renewable Energy, Government of India. (2023). *State-wise installed capacity of Renewable Power as on 31.08.2023.*

Petroleum Minister Hardeep S Puri. (2023, October 10). *India's energy demand will continue to provide fuel for future economic growth and is bound to grow exponentially in coming years.* PIB Delhi.

Polo López, C. S., & Frontini, F. (2013, September 23–25). Energy efficiency and renewable solar energy integration in heritage historic buildings. In *Proceedings of the international conference on solar heating and cooling for buildings and industry (SHC 2013)*.

Rudiany, N. P., Isnarti, R., Ramadhan, I., & Kusumawardhana, I. (2023). Aligning "energy culture" as part of sustainable tourism ideas. *IOP Conference Series: Earth and Environmental Science*, *1199*(1), 012036.

Times of India. (2016, July 11). Solar panels to help monuments tap free energy. *Times of India.*

Times of India. (2023, September 7). Sanchi becomes India's 1st solar city. *Times of India.*

VR, A. (2023, April 6). *The role of solar energy in supporting sustainable tourism and travel.* Arka360.

Chapter 12

The Role of Tourism in the Economic Development of Jammu and Kashmir Region

Raushan Kumar[1], *Niranjan Deo Pathak*[2] *and Shiv Swaroop Jha*[3]

Amity University, India

Abstract

Kashmir is widely recognised as a prominent tourist destination within the Himalayan region of India. The Valley is abundant with a diverse range of valuable tourism assets. In order to ensure the sustainable utilization of these tourism resources, the implementation of an appropriate tourism policy is necessary. The primary objective of this study is to analyse government policies pertaining to the expansion and advancement of tourism in the Kashmir region. Additionally, the study also seeks to evaluate the potential for tourism and the influx of tourists in Kashmir. The Government of India has developed a preliminary tourist policy, as indicated by the research findings. It also focuses on enhancing human resources and tourism infrastructure, ensuring the safety and security of tourists and promoting tourism education within the state. Furthermore, the government is diligently endeavouring to foster the growth of ecotourism and lesser known tourist locations through collaborative efforts with many relevant entities. This study utilises secondary data sources to analyse the economic implications of tourism in the region of Jammu and Kashmir. It aims to investigate several indicators of economic progress, including tourist arrivals, job creation, the state's gross domestic product (GDP), infrastructure development and regional advancement. In addition to the agricultural industry, the tourist

[1]https://orcid.org/0000-0003-4064-9480
[2]https://orcid.org/0000-0002-4653-6711
[3]https://orcid.org/0000-0003-4256-7243

Sustainable Tourism, Part B, 209–221

doi:10.1108/978-1-83608-326-920241012

sector has emerged as a prominent contributor to the economy, serving as a significant source of income and employment opportunities.

Keywords: Tourism; economic development; Jammu and Kashmir; gross domestic product; tourism asset

1. Introduction

On 5 August 2019, Prime Minister Narendra Modi and the Indian Government tried to address a prolonged 30-year insurgency by implementing a significant amendment to the constitution of the country. This amendment involved the revocation of autonomy provisions for the state of Jammu and Kashmir (J&K), the separation of J&K from the territory of Ladakh and the reclassification of both regions from statehood to union territories. After that period, there has been a notable increase of measures aimed at restricting political and civil liberties within the Kashmir Valley, which has garnered considerable attention from the international community. The sudden and independent alteration of Kashmir's status is a significant event in the region, comparable to the insurgency outbreak in 1989 or the nuclear tests conducted by India and Pakistan in 1998. Consequently, it is essential to conduct a comprehensive analysis of the factors that led India to this juncture and the potential trajectory of the region.

The maintenance of peace in the region of Kashmir holds significant importance for ensuring stability across the broader South Asian region. A mere half year before the Indian Government's decision to revoke Article 370 of the constitution, an act of suicide bombing targeted against Indian security forces in South Kashmir initiated a precarious interstate crisis. This incident led to a swift escalation between India and Pakistan, both of which had nuclear capabilities, causing significant concern among policymakers.

1.1 J&K: A Brief History

The presence of simmering tension, occasionally escalating, may be perceived as inconsequential by certain individuals; yet, it poses a substantial and imminent danger to regional stability. Kashmir, situated in the northwestern region of the Indian subcontinent, is sometimes referred to as the 'Switzerland of Asia' because to its mesmerising natural scenery. This Himalayan province has garnered such recognition. The present-day disputed region, which emerged as a consequence of the partition of British India, is currently partitioned between Pakistan and India, delineated by a de facto border known as the Line of Control (LOC). The historical background of the region indicates that J&K has seen significant unrest prior to the partition of the Indian subcontinent. Kashmir has not been governed autonomously by its indigenous population since it is ruled by the Mughal emperors during the historical period. The region saw successive periods of governance by Afghans (1753–1819), Sikhs (1819–1846) and subsequently by Dogras, who ruled the land through the Treaty of Amritsar in exchange for a payment of 7.5 million rupees (equivalent to 48,554 USD). Gulab Singh Dogra assumed the role of monarch of the region and governed it from 1846 until the partition of the subcontinent occurred (Ali et al., 2019).

The current source of animosity between the two nations can be attributed to the Kashmir issue. The ruler of Kashmir, Maharaja Hari Singh, was faced with the dilemma of choosing which country to align with. The individual in question expressed a desire to maintain autonomy, disregarding the stipulation outlined in the Independence Act of 1947 that mandated the valley's affiliation with one of the two nations prior to 15 August 1947. For a duration of about 2 months, the region of Kashmir maintained its autonomy until it was attacked by tribes coming from Pakistan. Consequently, the Maharaja of Kashmir chose to align with India, which caused dissatisfaction and annoyance on the part of Pakistan (Schofield, 2003, p. 7). Following the invasion, approximately one-third of the territory fell under the jurisdiction of Pakistan, while the remaining portion came under the control of India.

1.2 Economic Development

The basic principle of economic development is to develop the standard of living of economically excluded people and underdeveloped areas. Development economies also mean to increase the national income of the people and fostering the rate of development over population growth. Earlier, economic development stressed more about development in economic terms, but in the modern approach, economic development is more about economic well-being of the people.

Till the 1960s, the term economic development and economic growth were often used as the synonyms, but after increasing number of economist and their broad thinking, the term economic development was narrowly defined as the reduction of poverty, increasing per capita income and reducing inequality.

1.3 Demographic Profile of J&K

J&K represents the northernmost part of India. The overall land area of Jammu covers 222,236 square kilometres, with Pakistan occupying 78,114 square kilometres and China occupying 37,555 square kilometres. The remaining geographical region is partitioned into three distinct divisions, namely J&K and Ladakh. Jammu encompasses an extent of 26,293 square kilometres, while Kashmir spans 15,948 square kilometres. Ladakh, being the largest division, covers an expansive area of 59,146 square kilometres. Until October 2019, the region under consideration was officially recognized as a state, comprising three distinct administrative divisions, namely J&K and Ladakh. According to the 2011 census, the population of the state was 1,25,41,302. Following the implementation of the J&K Reorganization Act 2019, the region of J&K saw a significant transformation as it transitioned from being a state to being divided into two distinct centrally managed union territories. These areas are now known as the Union Territory of J&K and the Union Territory of Ladakh. According to the

2011 census, the population density of the state was recorded at 124 individuals per square kilometre. However, following the bifurcation, the population density of the Union Territory of J&K has increased to 290 individuals per square kilometre, while the Union Territory of Ladakh presently exhibits a population density of 4.6 individuals per square kilometre. The Union Territory of J&K is renowned for its plethora of tourist attractions, attracting a substantial number of visitors annually. The Jammu area is widely recognized for its reputation as a city renowned for its numerous temples. Every year, a significant number of pilgrims from various regions of India visit Jammu in order to worship Mata Vaishno Devi shrine in Katra. The sacred temple is situated on the Trikuta hills inside the Reasi district. The Kashmir division was renowned as a paradisiacal region, and it is said 'Switzerland of India' and its breathtaking natural landscapes and favourable climatic conditions, characterised by distinct seasons that captivate individuals worldwide.

1.4 Economic Profile of J&K

As stated earlier, the economy of J&K is primarily reliant on agriculture. The agricultural industry is a significant contributor to the state's economy, accounting for the employment of over 60% of the state's workforce (Economic Review, 2016). Due to the constraints posed by limited cultivable land and challenging agro-climatic conditions, the progress of agricultural practices has remained stagnant over the course of time. The expansion of industrialisation in the state is hindered by adverse geographic conditions such as hilly terrain, transportation bottlenecks and challenging climatic conditions that cannot be easily improved. However, the state possesses abundant natural resources including gypsum, limestone, coal, quartzite, lignite and forest resources. Therefore, industries have the potential to be implemented on a large scale in order to effectively utilise the available local resources. However, a significant portion of the state's natural resources, such as water, forests and minerals, have mostly remained undiscovered and underutilised (Gani & Labour-management Relations, 1991). The state's competitive advantage in these natural resources offers ample opportunity for the process of industrialisation. The state also possesses stunning landscapes, internationally recognized health resorts, captivating scenery, and true cultural heritage in crafting skills, hence occupying a unique position in the travel and handicraft business. The state provides a range of incentives that are available to industries in order to support their services. A network of financial institutions has emerged across the geographic expanse of the state. Furthermore, a comprehensive network of institutions has been established to facilitate and promote the advancement of enterprises within the state. Moreover, the current economic and ecological competitive situation in the state does not pose a significant threat to the robust industrial landscape. Nevertheless, the primary limitation hindering the favourable industrial situation in the state is to its challenging socio-cultural demands (Economic survey J&K, 2014–15).

The region of J&K possesses a substantial reservoir of both people and natural resources. However, due to the constraints imposed by the natural environment and geographical factors, establishing industries with a substantial capital base is challenging. Based on the data released by the Department of Industries and Commerce in J&K, the region is home to a total of 55,742 small-scale firms, which collectively provide employment opportunities to approximately 267,194 workers. The primary constituent elements within the region encompass food products, metal goods, wood and hosiery. The forests within the state provide significant quantities of raw materials to many industries, including high-quality timber for matchstick production, willow wood for the manufacturing of cricket bats and walnut trees for the art of wood carving. Industries such as cotton, silk, carpet weaving, and wood carving have demonstrated significant success in the state, contributing substantially to employment generation (Economic survey of J&K, 2014–15). Artistic creations hold a crucial role in the economic structure of the nation. The state possesses a vast heritage of artistic creations that are unparalleled. The handicraft industry has a significant role in generating wealth through both intra-state and inter-state trade, particularly in regions where large-scale industries are lacking. The importance of handcraft industries can be comprehended by the fact that this sector provides employment opportunities to almost 300,000 workers. The handicraft industry of the state is widely recognized on a global scale. Embroidered pashmina, Kani shawls and Raffal shawls are esteemed artistic creations that have gained international acclaim (Economic survey Jammu and Kashmir, 2014–15).

1.5 Types of Tourism in J&K

1.5.1 Religious

Religious tourism involves the act of visiting sacred sites with the intention of gaining inner tranquilly and spiritual fulfilment. Kashmir is home to various sacred sites associated with the Hindu, Muslim, Christian and Sikh religions. In addition to serving as spiritual and meditation centres, the pilgrim locations in Kashmir also offer a visually captivating experience. The phenomenon of pilgrimage tourism in the region of Kashmir is experiencing a significant surge in popularity. Annually, several religious adherents from all regions of the nation embark on pilgrimages to Kashmir, specifically to sites of religious significance such as Amarnath, Martand and Kheer Bhawani.

1.5.2 Wildlife Tourism

Wildlife tourism encompasses the act of travelling to certain regions with the purpose of observing and experiencing the indigenous plant and animal species in their undisturbed natural environments. The Kashmir Valley is surrounded by mountains on all sides. The notable animal attractions in this area comprise Dachingam National Park, Baltal Animal Sanctuary and Gulmarg Biosphere Reserve. Birds of different shapes and sizes can be found in the Kashmir Valley.

Nearly 287 bird species and 42 fish species call the Valley home. The Black-Necked Crane is an uncommon species of crane that can be observed in the Kashmir Valley, as well as in Tibet and China.

1.5.3 Cultural Tourism

Cultural tourism can be defined as a form of travel that aims to gain insights into the customs, traditions and historical background of a particular destination (Richards, 2007). Kashmir possesses a diverse and vibrant cultural heritage that is widely recognised and appreciated worldwide. The culture of Kashmir is characterised by the confluence of several ethnic groups with diverse religious beliefs. Hospitality constitutes a fundamental element of its cultural fabric. The people residing in the region of Kashmir exhibit a profound adherence to cultural values, evident via their various artistic expressions such as dance and music, observance of festivals, engagement in arts and crafts and the development of their culinary traditions.

1.5.4 Adventure Tourism

Adventure tourism is a distinctive form of outdoor activity that may be categorised as a recreational travel pursuit in which the presence of physical risk and hazard is an inherent and necessary component (Dar, 2014). Kashmir is widely regarded as an exceptional adventure location on a global scale. Indeed, this destination serves as a haven for enthusiasts of thrilling experiences, characterised by a multitude of physically demanding dangers and demands. The picturesque landscape of Kashmir provides an optimal environment for a variety of thrilling adventure pursuits, such as mountain climbing, trekking, climbing rocks, rafting, skiing, paragliding and other activities.

2. Objectives of the Study

- To analyse the impact on tourism and terrorism after the revocation of Article 370.
- To analyse the tourist inflow in J&K pre- and post- revocation of Article 370.

3. Research Methodology

The present research relies on secondary sources. The data have been gathered from multiple sources, including the Ministry of Tourism Government of India, J&K Tourism Development Corporation, Directorate of Tourism, Annual Economic Survey of J&K and World Travel and Tourism Council. Furthermore, the data have been gathered from reputed web sources such as publications, scholarly journals and newspapers.

4. Analysis and Discussion

4.1 Tourism in J&K: After Article 370

The J&K budget 2021–2022 stated that there is a surge in the rate of economic growth of 14.64% in the union territories of J&K, and with increase of economic growth, collection of tax revenue has also increased by 31%. After the statehood of J&K region, there is a sudden increase in the footfall of tourist in the valley. In the year 2021–2022, the total number of tourists that have visited to the valley are 1.88 crore.

Tourism has grown as a significant and prominent contributor to the state's economy. Until 1989, tourism constituted the primary economic endeavour inside the state. Nevertheless, the region experienced significant economic setbacks in the aftermath of the breakout of militancy during the previous two decades. The circumstances have had a significant impact on the economic stability and job prospects of the population. Nevertheless, tourism in the state has experienced a recent resurgence. The year 2011 has achieved a significant milestone by breaking all prior records in terms of the number of tourists and pilgrims arriving in the state. According to available data, the number of visitors who visited J&K in 2012 was approximately 13 million, positioning the region at the 17th spot in terms of tourist influx. In contrast, Andhra Pradesh claimed the top position among the prominent tourist destinations in India. Table 12.1 displays the data pertaining to the influx of tourists in the regions of J&K and Ladakh for the period spanning from 2014 to 2020.

4.2 Terrorism: After Article 370

The primary motivation behind the revocation of special status was the objective to eradicate the menace of terrorism. Based on the data presented by the Ministry of Home Affairs, there has been a significant decline in incidents related to terrorism after the repeal of Article 370 (Syed & Khan, 2021). According to the

Table 12.1. Projections of Tourists in J&K and Ladakh (in Lacs).

Year	Tourist in Lacs	Employment in Lacs
2014	147.43	22.45
2015	167.22	23.76
2016	178.43	25.45
2017	18.67	27.64
2018	196.43	29.76
2019	211.23	31.67
2020	243.73	34.85

Source: Directorate of Kashmir (Ministry of Tourism, Government of India, 2020).

data provided by the Ministry of Home Affairs in the Rajya Sabha, there has been a 59% reduction in the frequency of terrorist incidents in the year 2020 as compared to the preceding year, 2019. In contrast to the corresponding time in 2020, there was a further decline of 32% in the frequency of events observed up until June 2021. Based on the data, it can be observed that a total of 56 individuals have lost their lives in incidents associated with terrorism since August 2019. Conversely, no civilian casualties have been reported in scenarios involving law enforcement and related incidents during this comparable timeframe. Nevertheless, considering the recent drone strikes on the Indian Air Force (IAF) station located at the Jammu airport, there has been a growing sense of apprehension regarding the issue of cross-border terrorism. Unmanned aerial vehicles were employed for the inaugural instance in this assault, so instigating concerns regarding the potential occurrence of increasingly advanced and intricate attacks in subsequent instances. Terrorism emerged as a significant challenge for the region of J&K, which was previously recognized as an independent state. Pakistani terrorists successfully infiltrated J&K, thereby facilitating the proliferation of their malicious endeavours across the nation. This was made possible through the collaboration of extremist Kashmiris and separatist leaders. Nevertheless, since the abrogation of Article 370, there has been a notable decline in the frequency of terrorist incidents within the recently established union territories.

4.3 Recuperation of Kashmiri Hindus

Based on a 1990 data, a total of 44,168 households originating from Kashmir had relocated from the Valley because of security apprehensions that emerged during the 1990s. Among the listed individuals, a total of 39,785 migrants of Hindus are included. Based on the official response provided by the government in the Rajya Sabha in March 2021, it has been reported that a total of 3,841 Kashmiri migrant youth have repatriated to the Kashmir region in recent years. These individuals have actively participated in various districts, availing themselves of the rehabilitation package offered by the Prime Minister (Farooq & Javaid, 2020). According to the available information, individuals were appointed to the positions through the programme in April 2021, specifically in the year 1997. The administration has announced its commitment to developing a comprehensive strategy aimed at facilitating housing solutions for migrants who have repatriated to the region of Kashmir (Ahmad, 2019).

4.4 Industrialisation: After the Article 370

According to the report to the Rajya Sabha in March 2021, a total of 460 Memorandums of Understanding (MoUs) were signed with potential investors in the newly constituted Union Territory of J&K. These MoUs have an estimated value of Rs 23,154.20 crore. According to the statement, all the primary endeavours of the Modi administration, which encompass various structures centred on benefiting certain groups, are currently implemented throughout the

Union Territory. The Central Administration has recently approved a Business Expansion Scheme worth Rs 29,500 for the state of J&K. This scheme aims to promote industrial activity and support the growth of existing businesses (Mohydin, 2021). Furthermore, it is noteworthy to mention that in August 2020, the administration of J&K granted approval for a business stimulus package amounting to Rs 1,422.50 crore.

5. Reasons for Sluggish Economic Development in J&K

5.1 Historical Context

The historical and political conflict in J&K has played a pivotal role in hampering economic development (Ganguly, 1996). The region has witnessed political instability for decades, which has impeded economic development. Frequent changes in government, separatist movements and security concerns have discouraged businesses from investing in the region.

5.2 Agricultural Challenges

According to the Economic Survey of J&K. (2020), the region faces agricultural challenges, including land fragmentation and limited modernisation, which hinder agricultural productivity. Despite the importance of agriculture, the sector faces challenges such as land fragmentation, inadequate irrigation facilities and limited access to modern farming practices, which limit agricultural productivity.

5.3 Economic Diversification

According to Gani & Khan (2005), lack of diversification and the economy's heavy reliance on agriculture and tourism necessitates diversification to enhance economic resilience.

5.4 Tourism Potential

J&K has immense tourism potential, but security concerns and inadequate infrastructure have hindered its growth (Mushtaq, 2014). J&K's tourism potential has been hampered by security concerns, lack of infrastructure and communication issues. The region has great potential for tourism, but it has not been fully realised due to these challenges.

5.5 Infrastructure and Connectivity

Infrastructure deficiency and inadequate infrastructure, such as road networks, power supply and connectivity, poses significant challenges to industrialisation and trade (Ministry of Road Transport and Highways, Government of India, 2019).

5.6 Entrepreneurship and Skill Development

Limited Access to Credit: Limited access to credit and financial services hampers the growth of small- and medium-sized enterprises (SMEs) (SLBC, Jammu & Kashmir, 2021–2022).

Skill Development: The lack of skill development programmes and quality educational institutions affects employability and entrepreneurship. The lack of skill development programs and quality educational institutions has hindered the employability of the local population. This, in turn, impacts the region's economic growth by limiting the availability of a skilled workforce (State industrial profile, 2016).

5.7 Security Concerns

Security Issues: Persistent security concerns, including insurgencies and border tensions, deter private investment and have a direct impact on the economy (Reddy & Reddy, 2018). The persistent security concerns in J&K, including insurgencies and border tensions with Pakistan, have deterred private investment and economic growth. The security situation has a direct impact on tourism, agriculture and trade.

5.8 Geographical Isolation

According to Sinha's (2017) geographical challenges, J&K's difficult terrain and limited connectivity contribute to its economic isolation. The financial sector in J&K has been relatively underdeveloped, and access to credit for businesses and entrepreneurs has been limited. This has restricted the growth of small- and medium-sized enterprises (SMEs) and entrepreneurship.

5.9 Government Policies and Interventions

Government Initiatives: Analyse the impact of various government policies and initiatives aimed at addressing economic challenges in the region. Articles 370 and 35A of the Indian Constitution gave the state of J&K extra autonomy. In August 2019, the Indian Government took those away. This change led to increased central government control and administration. While the move was aimed at integrating the region into the Indian Union, it initially caused uncertainty and unrest. The aforementioned factor had an adverse effect on the overall business environment and investment climate (Government of Jammu and Kashmir, 2019).

6. Suggestions

6.1 Robust Cooperation

The development of a robust cooperation between the government and public agencies, with a shared focus on the welfare of the general populace, is crucial for fostering the economic revitalisation of J&K using contemporary approaches.

6.2 Building Tourism Infrastructure

It is important to ensure that infrastructure tourism-related, encompassing transportation, travel services, restaurants, recreation and entertainment activities, as well as emergency services, prioritises safety and security measures. An online medium should be created to facilitate the exchange of experiences between visitors to the valley and anyone interested in visiting the area.

6.3 International Visits

The government should develop approaches aimed at attracting affluent and prominent people to visit the region of Kashmir. In order to achieve this objective, it is needed to engage embassies and access international media platforms.

6.4 Sustainable Tourism Development

Recent economic development in J&K progressive, but Tourism can be very good for a J&K economy, but they can also be very bad for it. The rapid expansion of tourism can result in excessive development, which places significant pressure on natural resources, ecosystems and landscapes (Kaul & Gupta, 2009). More tourists can pollute the environment by doing things like driving and making more trash. This can affect the quality of the air and water. Unregulated tourism can result in the commercialisation of indigenous cultures, diminishing traditional customs and principles in order to meet the demands of tourists. The presence of a significant influx of tourists might disturb the local populations and their customary lifestyles, leading to strained relations and disputes between the residents and the visitors. Tourism can result in over-utilisation of natural resources, including water, energy and land, especially in environmentally vulnerable regions. To mitigate these adverse effects, it is imperative for the J&K Government to embrace sustainable tourism practices, execute responsible tourism management techniques and participate in community-based tourism planning. Ensuring a harmonious equilibrium between the economic advantages of tourism and the imperative of safeguarding the environment and preserving cultural heritage is of utmost importance for the sustained welfare of both the J&K tourist destination and its inhabitants (Chettiparamb & Kokkranikal, 2012; Menon et al.,2021; Yadav et al., 2018).

7. Conclusion

History provides ample evidence that in any instance of significant violence, the economy of the affected region inevitably suffers as a primary consequence. This is also the situation for the region of J&K. The aforementioned study indicates that the economy of J&K has experienced a precipitous decline as a result of significant political turmoil. This resulted in a severe slump in the economy. Because of the disruption, the development of the state has been negatively impacted, and private businesses have been dissuaded from making investments as a result, which has created many obstacles in the path to economic prosperity. Since the disruption began three decades ago, the state has already incurred enormous costs as a result of it. It is conceivable for J&K to prosper in the near future if all sides are involved in the process of participating in a better mutual solution, which should be developed so that it may be implemented. The regular people of the state have been waiting for quite some time for this to bring about the peace and prosperity that they have been hoping for.

References

Ahmad, J. (2019, August 11). Will the removal of Article 370 bring peace? *The Economic Times*. https://economictimes.indiatimes.com/news/polities-and-nation/swill-the-removal-of-article-370-bring-peacc/articleshow/70622110.cms?utm_source=contentofinterest&cutm_medium=text&utm_campaign=cppst

Ali, R. N., Khan, A. A., & Iqbal, A. (2019). Political consciousness of the Muslims in Jammu and Kashmir state 1846–1947. *Journal of the Research Society of Pakistan*, *56*(1), 15.

Chettiparamb, A., & Kokkranikal, J. (2012). Responsible tourism and sustainability: The case of Kumarakom in Kerala, India. *Journal of Policy Research in Tourism, Leisure and Events*, *4*(3), 302–326.

Dar, H. (2014). Potentials and problems of adventure tourism: A study of Kashmir Valley. *Abhinav International Monthly Refereed Journal of Research in Management & Technology*, *3*(9).

Directorate of Economics and Statistics, Jammu and Kashmir. (2020). *Economic survey of Jammu and Kashmir*. https://ecostatjk.nic.in/pdf/publications/ecosurvey/Economic%20Survey-2023.pdf

Economic survey Jammu and Kashmir. (2015–16). https://ecostatjk.nic.in/pdf/publications/ecosurvey/2014-15.pdf

Economic survey Jammu and Kashmir. (2015–16). https://ecostatjk.nic.in/pdf/publications/ecosurvey/2016.pdf

Farooq, M., & Javaid, U. (2020). Suspension of Article 370: Assessment of Modis Kashmir masterstroke under Hindutva ideology. *Global Political Review*, *5*(1), 1–8.

Ganguly, S. (1996). Explaining the Kashmir insurgency: Political mobilization and institutional decay. *International Security*, *21*(2).

Gani, A., & Labour-management Relations. (1991). *A study of textile industry in Jammu and Kashmir*. Concept Publishing Company.

Gani, A., & Khan, S. (2005). *Diversification of the economy of Jammu and Kashmir*. Institute of Kashmir Studies, University of Kashmir.

Government of Jammu and Kashmir. (2019). *Jammu and Kashmir Reorganization Act*. https://igr.jk.gov.in/files/J&K%20Reorganisation%20Act,%202019.pdf

Kaul, H., & Gupta, S. (2009). Sustainable tourism in India. *Worldwide Hospitality and Tourism Themes*, *1*(1), 12–18.

Menon, S., Bhatt, S., & Sharma, S. (2021). A study on envisioning Indian tourism – Through cultural tourism and sustainable digitalization. *Cogent Social Sciences*, *7*(1), 1903149.

Ministry of Road Transport and Highways, Government of India. (2019). *Report on infrastructure development in Jammu and Kashmir*. https://morth.nic.in/sites/default/files/Ministry%20Annual%20Report_2019-20.pdf

Ministry of Tourism, Government of India. (2020). *20 year perspective plan for sustainable development of tourism in Jammu and Kashmir*. Santek Consultants Private Limited.

Mohydin, R. (2021). The Kashmir tinderbox: International media coverage of the revocation of Article 370. *The Political Economy of Communication*, *8*(2).

Mushtaq, K. (2014). Impact of terrorism on economic development in Kashmir. *International Journal of Business and Economics Research*, *3*(2).

Reddy, G. J., & Reddy, G. V. K. (2018). Cross-border terrorism and insurgencies: Implications for India's internal security. *Sinews of India's internal security in 21st century*, 29.

Richards, G. (2007). *Cultural tourism: Global and local perspectives*. Psychology Press.

Schofield, V. (2003). *Kashmir in conflict*. I.B Tauris & Co Ltd.

Sinha, R. K. (2017). *Jammu & Kashmir: Past, present and future*. Indian Council for Research on International Economic Relations (ICRIER).

State industrial profile. (2016). *Micro, small and medium enterprises*. https://dcmsme.gov.in/old/dips/state_wise_profile_16-17/NEW%20SIP%20FOR%20PDF.pdf

State Level Bankers' Committee, Jammu and Kashmir. (2021–22). *Annual credit plan for Jammu & Kashmir*. https://www.jkslbc.com/

Syed, K. H., & Khan, I. (2021). Articles 370 and 371 of the Indian Constitution in the Context of Kashmir. *Sir Syed Journal of Education & Social Research*, *4*(1), 286–294. https://doi.org/10.36902/sjesr-vol4-iss1-2021(286-294)

Yadav, N., Sahu, N. C., Sahoo, D., & Yadav, D. K. (2018). Analysis of barriers to sustainable tourism management in a protected area: A case from India. *Benchmarking: An International Journal*, *25*(6), 1956–1976.

Chapter 13

Development of Ecotourism in Himachal Pradesh: A Case Study of Spiti Valley

Alka Maheshwari[1] *and Aruditya Jasrotia*[2]

Amity University, India

Abstract

Spiti Valley is a home to temples, untouched natural landscapes and is enriched in flora and fauna. This makes Spiti Valley an emerging ecotourism destination. The prime objective of this study is to analyse the development of ecotourism in the state of Himachal Pradesh, India, particularly in the region of Spiti Valley, Himachal Pradesh. This study provides an opportunity to examine the level of awareness about ecotourism among tourists. In order to achieve the objectives of the study quantitative methodology has been adopted. The data were collected through a questionnaire. The findings of the study indicate towards the awareness among the tourists regarding the environment and how everyone is playing their parts to conserve and protect the environment. The study also depicts the economic benefits of ecotourism to the region and how it can generate employment for the local people directly or indirectly.

Keywords: Ecotourism; tourism development; Spiti; Himachal Pradesh; sustainable tourism

1. Introduction

The necessity for preserving the environment has been globally recognised in the past 40 years. In the 1970s, the environmental movement named 'Promoting the preservation of the earth's limited natural resources' gained momentum. The first national policy for promoting the preservation of the wildlife in India was the

[1]https://orcid.org/0000-0002-5204-1686
[2]https://orcid.org/0000-0002-8274-5609

Sustainable Tourism, Part B, 223–232

doi:10.1108/978-1-83608-326-920241013

Wildlife Protection Act of 1972. The idea of ecotourism emerged within the tourism sector as a means of promoting greater environmental consciousness. Sustainable tourism aims to minimise the adverse impacts and enhances the positive effects of various forms and activities of tourism on local communities, the environment, cultural and natural heritage and inclusive economic growth (Brush & King, 1979). The idea of ecotourism is applicable to all forms of tourism. The most notable partnership of ecotourism tourism is with nature-based tourism (Chaudhary et al., 2019).

Due to the environmental movement, the 1980s witnessed an increase in the number of ecologists and biologists travelling to remote locations, and often biodiverse locations to gain knowledge about unexplored areas of the globe. This targeted the commercialisation of nature-based tourism (Mishra et al., 2003). In combination with an elevated awareness of the consequences of the environmental impacts, nature-based tourism shifted towards an ethic of sustainability, and therefore the term 'ecotourism' emerged to define this new form of tourism. Since its beginning, ecotourism has developed into one of the rapidly growing sectors of the tourism industry (WTTC, 2018). In the year 2000, Bjork described ecotourism as 'the sustainable travel to natural areas, that is environmentally conservative while maintaining the local culture and contributing to the well-being of the host community'. From this definition, it can be observed that ecotourism is a source of creating environmental awareness, cultural exchange and ecological preservation in a sustainable manner (Honey, 2008; Kanjihal, 2008) Ecotourism, alternatively referred to as ecological tourism, involves conscientious travel to pristine, sensitive and typically preserved regions with the objective of minimising impact and often limiting scale. Its purpose is to educate travelers, allocate funds for ecological conservation, directly contribute to the economic progress and political empowerment of local communities and foster appreciation for diverse cultures and human rights. Ecotourism is considered essential by all those who take part in it, so that the future generations may witness those characteristics of the environment which are relatively untouched by human interference (Mishra, 2000). Unrestrained tourism, devoid of sustainable practices, is disrupting the delicate balance of ecosystems and endangering biodiversity in Himachal Pradesh. To combat this destruction, tourism activities must be transformed into eco-friendly endeavours. Ecotourism has gained significant traction in recent years, particularly in developing nations (Bhardwaj, 2022). Currently, there is a need to encourage ecotourism in Himachal Pradesh to safeguard and maintain the richness of both natural and cultural surroundings while actively engaging the local community. Emphasising community-based tourism is now recognised as a sustainable approach, as it directly includes and supports local communities, contributing to the conservation of biodiversity (Panchani, 2018; Verma & Thakur, 2015). This chapter aims to fill this gap by analysing the level of awareness among people and by giving valuable suggestions for developing ecotourism in Spiti.

2. Ecotourism Policy of Himachal Pradesh (2016)

During 2005, an Ecotourism Policy was framed by the Forest Department of Himachal Pradesh in order to allow the tourists to witness this natural heritage (Blanchard & Higgins-Desbiolles, 2013; Fisher &Turner, 2008). However, amendments were required in the policy primarily because of the systematic necessities of acquiring clearance of Eco-tourism destinations under Forest Conservation Act, 1980 on the basis of the guidelines of the Ministry of Environment, Forest & Climate Change (MoEF & CC), Government of India. Few amendments in the factors of the developing understanding of Eco-Tourism ideas and principles have been called for in the existing policy document (Bartelmus, 2009).

The aim of the Revised Eco-Tourism Policy of 2016 is to bring the wild and untouched ecosystems of Himachal Pradesh more proximate to those who visit the place and simultaneously ensures sufficient systems and protective measures for the protecting and conserving the natural resources. By the involvement of the locals, the policy aims to help in increasing the income opportunities and their participation in creating awareness, protecting and conserving. It will also help in anticipating the creation of income which could be used in proper up keeping and maintaining the nature. It also promotes larger understanding and respect for the natural and cultural heritage (Mowforth & Munt, 1998). The basis of the policy is the belief that involving the locals in ecotourism would assist their financial needs and in consequence will create a support for them in the conservation of ecology, local culture and the environment (Montaguti & Mingotto, 2015).

According to this policy, visitors, being important stakeholders, possess an important function of responsibly travelling to natural areas and making available financial contribution for conservation of natural heritage and empowerment of local people. On the other hand, local communities, being another major stakeholder, possess the prime function of providing positive experiences for the visitors.

2.1 Current State of Eco-Tourism in Himachal Pradesh

Himachal Pradesh holds significant potential to contribute to the nation's economic growth and set a new trajectory. The state's tourism opportunities are surpassed only by its hydropower potential. It offers a range of choices for wildlife enthusiasts and outdoor enthusiasts alike. Thanks to its diverse geography and environment, it emerges as a paradise for climbers, trekkers and other adventure seekers. The region provides ample space for activities such as trekking, paragliding, mountaineering, rock climbing, heli-skiing and water-based pursuits like golf, water skiing, fishing, river rafting and more. Until recently, many travelers were unaware of these sports, but even now, it continues to attract enthusiastic athletes from around the world (Chandra, 2018). Prime challenges faced by the state of Himachal include poverty and environmental issues, currently ranking among the most pressing concerns. Small towns and villages often resort to utilising local natural resources, including forests, for livelihoods, contributing to

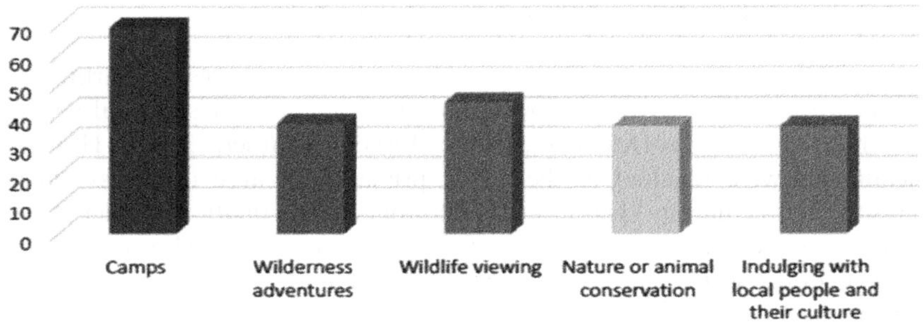

Fig. 13.1. Tourism Activities. *Source:* Data Analysis.

deforestation. Consequently, environmental protection tends to be overlooked in these areas (Nag, 2018). Recognising this dilemma, 'ecotourism' is gaining traction among environmentalists as a strategy to simultaneously promote environmental conservation and economic prosperity. Tourism, simply defined as 'the activity of travelling for enjoyment', becomes a potential avenue for achieving this dual purpose. The tourism industry in the state exerts socio-economic and environmental influences. The significant challenge lies in mountain tourism, where substantial efforts are underway to develop infrastructure for tourism promotion, accompanied by assessments of the tourism carrying capacity. Ecosphere is a thriving organisation in Spiti which is responsible and sustainable ecotourism. Ecosphere's approach seamlessly integrates economic empowerment, conservation and sustainable development, paving the way for a harmonious future in the region. Remarkably, Ecosphere embarked on its journey when the idea of homestays and their potential revenue was still a foreign concept to many. Consequently, Ecosphere's growth stands as a remarkable testament to the transformative power of innovation and its ability to shape the future.

3. Ecosphere of Spiti

Ecosphere is a socially active firm and is a joint attempt between the experts from varied professions with a broad range of expertise and learning, adequately spreading over the extension from the basic to the specialty and the local community of Spiti Valley (Ahmadi et al., 2018). The emphasis of the ecosphere is on creating sustainable livelihoods that are connected to cultural and natural conservation (Sharma & Pradhan, 2011). The basis of the establishment of this enterprise is the passion of the team members of the Ecosphere for mountain travel, conservation and adventure (Khah et al., 2011). The dedication of ecosphere is not only limited towards developing the people, place and resources of Spiti Valley, but also to the external world with whom they aspire to share the heritage enriched in cultural and natural aspects and the traditional knowledge of reviving and health of the Spiti Valley. Tourism is a feasible income generating

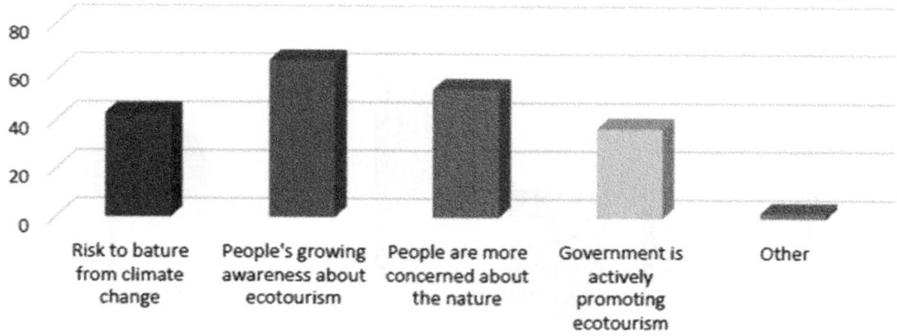

Fig. 13.2. Reason for Increase in Ecotourism Activities. *Source:* Data Analysis.

option for the local community in Spiti Valley because of its rich cultural and natural heritage (Boley & Green, 2015; Murali et al., 2017).

The Ecosphere operates a crucial greenhouse project in Spiti Valley, addressing a significant need in the region where green vegetables are entirely absent. The delivery of supplies to the valley, mainly sourced from neighbouring cities like Shimla and Manali, which takes approximately 2 days. This transportation not only increases the cost of vegetables but also contributes significantly to their carbon footprint. Additionally, during the 6-month-long winter period, when access to neighbouring regions is cut off, green vegetables become unavailable. Ecosphere's developed greenhouses play a pivotal role by allowing locals to cultivate vegetables not only during the summer but also throughout the winter, even in temperatures as low as −25 degrees Celsius. These greenhouses contribute to the improvement of local health, a reduction in emissions and the provision of an extra source of income for the residents. The tourists can also voluntarily take part in the actual construction of these greenhouses and can assist the local people in cultivating their own vegetables.

4. Research Methodology

This study is dedicated towards analysing the development of Ecotourism in Spiti Valley, Himachal Pradesh, India. This research employs a combination of primary and secondary sources to gather data. The primary data was acquired using a meticulously crafted, pre-structured questionnaire schedule designed specifically for information collection. The survey was conducted within a particular section of the society that has taken part in ecotourism activity and that has possibly travelled to Spiti Valley. The data were analysed with the help of the Google forms which provided a statistical representation of the responses as shown in Figs. 13.1–13.4. The secondary data was collected from books, journals, research papers, articles, magazines published by different authors, newspapers and websites. The literature search adopted a broad approach, encompassing a wide range

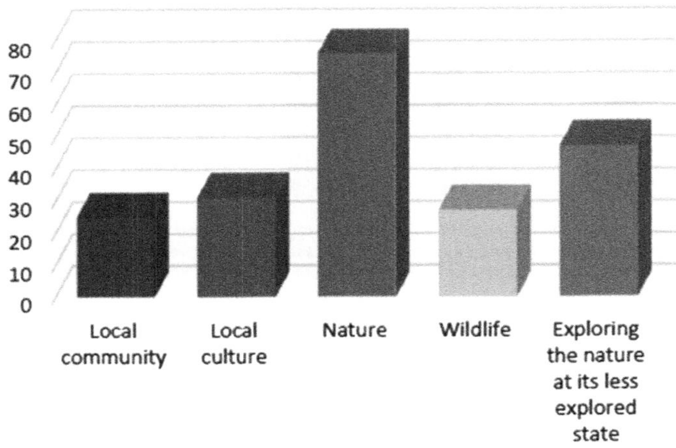

Fig. 13.3. Purpose of Visit. *Source:* Data Analysis.

of academic databases and relevant sources. To ensure the search's effectiveness, the language was restricted to English to maintain consistency and clarity. The search utilised a broad range of keywords encompassing terms associated with tourism, ecotourism, sustainable tourism, forest, Himalayas, conservation, environment, culture and their synonymous counterparts. This comprehensive approach guaranteed the identification of a diverse and all-encompassing collection of pertinent studies. This multifaceted approach ensured that a thorough and inclusive collection of relevant studies was identified.

5. Data Analysis and Findings

It was found in the study that out of all the respondents, 69 responded that they have been involved in camping. After camps, people like to be involved in wildlife viewing, i.e. visiting national parks, bird sanctuaries, etc. Apart from this, 35.9% share in the graph was grabbed by wilderness viewing as 37 people said that they have somewhere been involved in wilderness viewing. An equal number of responses were received for nature or animal conservation and indulging with local people and their culture and both of these activities' share was 35%. Local culture is one of the prime aspects of ecotourism and therefore more people should indulge themselves with local people and local culture of any ecotourism destination they are visiting.

Moreover, 63.1% of the respondents believe that the reason for increase in ecotourism in Spiti Valley is 'People's growing awareness about ecotourism'. Around 56% of the respondents believe that people are more concerned about the nature, and they don't want to harm it by practicing any tourism activity that is not in the favour of the nature. Around 42% of the respondents believe that people are aware about the risk to nature due to climate change and hence they don't want to contribute more to climate change or global warming and hence

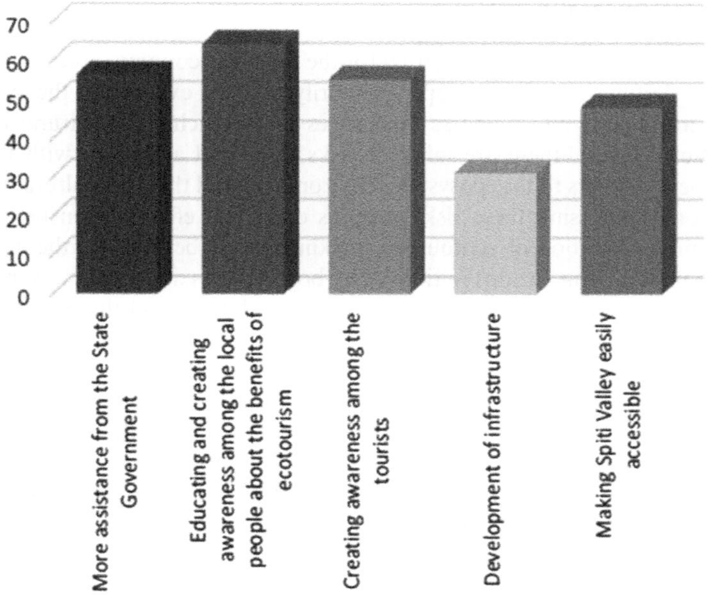

Fig. 13.4. Opinion for Promotion of Ecotourism. *Source:* Data Analysis.

ecotourism activities has increased as they don't possess any risk to nature. Only 36% of respondents believe that government is promoting ecotourism in Spiti Valley.

It is observed that the prime purpose of people visiting Spiti Valley is Nature as 73.8% people responded with the same. After nature, 45.6% responded that their purpose of visit was exploring nature at its untouched or less explored state. It is seen that local community, local culture and wildlife is not the prime purpose for tourists visiting Spiti Valley as only 24.3%, 30.1% and 26.2% responded, respectively. Tourists believe that nature in Spiti Valley is untouched and is at its original form; therefore, they make nature their prime purpose of visit and the rest secondary.

As the graph depicts, majority (62%) of respondents believe that to promote ecotourism in Spiti Valley, it is necessary to educate the local people and to create awareness among them regarding the benefits of ecotourism. Fifty-four percent of the respondents are of the view that more assistance should be provided by the State Government whereas 53% believe that it is important to create awareness among the tourists. Forty-seven percent of the respondents also believe that Spiti Valley should be made easily accessible so that the number of tourist footfalls may rise. only 30% of the respondents believe that there is a need to develop infrastructure in Spiti Valley.

6. Research Outcomes

Ecotourism holds significant potential for the region's economy, given its status as a tourist destination where people primarily visit to experience the beauty of various attractions and immerse themselves in the enchanting natural environment. Nevertheless, there are inherent risks that need attention, with the most serious being threats to the ecosystem, environment and the sociocultural fabric of the region. Addressing these risks requires concerted efforts to ensure the successful implementation of ecotourism, maximising its benefits for the area. This involves undertaking various initiatives by both visitors and locals to support the creation of economic opportunities for local populations and the preservation of the environment. Local culture is one of the prime aspects of ecotourism and therefore more people should indulge themselves with local people and local culture of any ecotourism destination they are visiting. Tourists believe that nature in Spiti Valley is untouched and is at its original form; therefore, they make nature their prime purpose of visit and the rest secondary. It is necessary to educate the local people and to create awareness among them regarding the benefits of ecotourism. Encouraging tourists to opt for eco-friendly transportation options can not only reduce the environmental impact of tourism but can also enhance their overall experience. Cycling and electric vehicles allow tourists to explore destinations at a leisurely pace, immersing themselves in the local culture and natural beauty. Moreover, cultural and heritage tourism can play a pivotal role in sustainable tourism within Spiti Valley, as the region is home to an array of culturally and historically significant sites that hold immense value for both local communities and the global community. The funds allocated for ecotourism should be employed in the conservation of biodiversity and in conducting awareness programs for society. The proceeds generated can be redirected to support additional facets of sustainable development, including tree planting, waste recycling and promoting ecological awareness (Panchani, 2018). In an endeavour to foster responsible tourism and safeguard the preservation of cultural and heritage assets, several cultural and heritage tourism initiatives have been implemented in Spiti Valley. Nestled amidst the awe-inspiring Himalayas, Spiti Valley stands as a shining exemplar of sustainable tourism in India. This captivating region, renowned for its rugged mountains, serene landscapes and rich cultural heritage, has wholeheartedly embraced sustainable practices, ensuring the preservation of its natural splendour while empowering its local communities.

7. Conclusion

Ecotourism is a type of tourism that emphasises exploration of local cultures, engaging in wilderness adventures, participating in volunteer activities, acquiring new lifestyles and fostering personal development. It is commonly characterised as travel to locations where the main attractions are the native flora, fauna and cultural heritage. Responsible ecotourism involves engaging in activities that minimise the negative impacts of conventional tourism on the natural environment and contribute to the preservation of the cultural identity of local communities. The analysis above indicates that the place has vast potential to develop as an ecotourism destination. There are certain issues, but if these issues are taken care of then, Spiti Valley can be developed as an attractive and most preferred destination for ecotourism. The issues that need to be taken care of are accessibility, infrastructure and quality of local service. This study also indicates that people are now pretty much aware about ecotourism and the benefits of ecotourism to the local people, the destination and the tourists themselves, and hence they are shifting from traditional form of tourism to ecotourism.

The study also indicates towards the awareness among the tourists regarding the environment and how everyone is playing their parts to conserve and protect the environment. The study also depicts the economic benefits of ecotourism to the region and how it can generate employment for the local people directly or indirectly. The Ecotourism Policy of Himachal Pradesh Government is also understood through this study, and which fulfils the objective of Role of State Government in promotion of ecotourism in Spiti Valley. Ecosphere Society of Spiti is playing a major role in promoting and conducting ecotourism in Spiti Valley. It is through the efforts of Ecosphere's several initiatives that the tourists and the local people are getting along with each other so well. Be it construction of greenhouses, living life like monks or staying in interactive homestays, ecosphere has made it easier for the local people to welcome the tourists and for the tourists to adapt themselves in the local culture.

References

Ahmadi, M., Khodadadi, M., & Shahabi, H. (2018). Planning for ecotourism in the protected area of Manesht and Ghelarang, Ilam Province, Iran. *Journal of Quality Assurance in Hospitality & Tourism*, *19*(2), 243–268.

Bartelmus, P. (2009). The cost of natural capital consumption: Accounting for a sustainable world economy. *Ecological Economics*, *68*(6), 1850–1857.

Bhardwaj, R. L. (2022). Role of ecotourism in sustainable development of Himalayas: A case study of Himachal Pradesh. *Rethinking Himalaya: Its Scope and Protection*, *109*.

Blanchard, L. A., & Higgins-Desbiolles, F. (Eds.). (2013). *Peace through tourism: Promoting human security through international citizenship*. Routledge.

Boley, B. B., & Green, G. T. (2015). Ecotourism and natural resource conservation: The 'potential' for a sustainable symbiotic relationship. *Journal of Ecotourism*, *15*(1), 36–50.

Brush, J. E., & King, A. D. (1979). Colonial urban development: Culture, social power and environment. *Journal of Asian Studies*, *38*(3), 605–607.

Chandra, S. (2018). Tourism in Himachal Pradesh – A case study of tourists attitude regarding basic facilities. http://www.ijiras.com/2017/Vol_4-Issue_6/paper_10.pdf

Chaudhary, S., Kumar, A., & Negi, M. (2019). A geospatial appraisal of Garhwal Himalayan bio-geodiversity and its ecotourism potentials. *International Journal of Research and Analytical Reviews*, *6*(1), 911–926.

Fisher, B., & Turner, R. K. (2008). Ecosystem services: Classification for valuation. *Biological Conservation*, *141*(5), 1167–1169.

Honey, M. (2008). *Ecotourism and sustainable development: Who owns paradise?* Island Press.

Kanjihal, G. (2008). An insight into eco-tourism in India. http://www.gourkanjilal.com/new-trends/eco-tourism-india.htm. Accessed on January 21, 2022.

Khah, S. A., Rao, R. J., & Wani, K. A. (2011). Ecotourism and impact of conventional tourism on fragile ecosystems. *International Journal of Science and Nature*, *2*(3), 432–442. ISSN 2229 – 6441.

Mishra, C. (2000). Socioeconomic transition and wildlife conservation in the Indian Trans-Himalaya. *Journal of the Bombay Natural History Society*, *97*(1), 25–32.

Mishra, C., Allen, P., McCarthy, T. O. M., Madhusudan, M. D., Bayarjargal, A., & Prins, H. H. (2003). The role of incentive programs in conserving the snow leopard. *Conservation Biology*, *17*(6), 1512–1520.

Montaguti, F., & Mingotto, E. (2015). Ecotourism in natural parks: An assured sustainable success? Tourist behavior, attractiveness and sustainable development issues in two Italian parks. *Tourism Planning & Development*, *12*(1), 99–110.

Mowforth, M., & Munt, I. (1998). *Tourism and sustainability: New tourism in the third world.* Routledge.

Murali, R., Redpath, S., & Mishra, C. (2017). The value of ecosystem services in the high altitude Spiti Valley, Indian Trans-Himalaya. *Ecosystem Services*, *28*, 115–123.

Nag, A. (2018). A study of tourism industry of Himachal Pradesh with special reference to ecotourism. *Sumedha Journal of Management*, *7*(1), 85–106.

Panchani, M. (2018). Challenges of biodiversity conservation & ecotourism in HP. *Voyager*, *11*(2), 5–15.

Sharma, P., & Pradhan, M. K. (2011). Eco-tourism as a strategy in supporting local livelihoods and protecting commons through community based institution. In *Sustaining Commons: Sustaining Our Future, the Thirteenth Biennial Conference of the International Association for the Study of the Commons*. https://hdl.handle.net/10535/7113

Verma, V., & Thakur, S. S. (2015). Analytical study of eco tourism to the tourist place by creating awareness case study of (Shimla). *AGU International Journal of Professional Studies & Research*, *1*(1), 8–17.

WTTC. (2018). *Travel and tourism economic impact 2018 world.* World Travel & Tourism Council. https://www.wttc.org/-/media/files/reports/economic-impact.2018/world2018.pdf. Accessed on November 6, 2018.

Chapter 14

Subsidy as a Means to Promote Sustainable Tourism for the Development of Economic Growth: A Critical Study

Parikshita Khatua[1] *and Liji Panda*[2]

KISS DU University, India

Abstract

The tourism industry plays a pivotal role in fostering economic growth, but its sustainability is crucial for long-term benefits. This study critically examines the effectiveness of subsidies as a strategic tool to promote sustainable tourism and contribute to economic development. The research explores various dimensions of subsidies and their impact on the triple bottom line – economic, social and environmental aspects of tourism. The study employs a comprehensive review of existing literature, case studies and empirical data to assess the outcomes of subsidy programs in different regions. By adopting a critical lens, the research evaluates the strengths and weaknesses of subsidy initiatives, considering their influence on the overall sustainability of tourism destinations. Key focus areas include the analysis of government policies, financial incentives and support mechanisms designed to encourage sustainable practices within the tourism sector. The study also investigates the role of subsidies in enhancing community engagement, preserving cultural heritage and mitigating environmental impacts associated with tourism activities. Moreover, the research scrutinises the potential challenges and unintended consequences linked to subsidy implementation. This critical assessment aims to identify gaps in current subsidy strategies and provide recommendations for policymakers, stakeholders and practitioners to refine and optimise subsidy programs for sustainable tourism development. The findings of this study contribute to the ongoing discourse on the intersection of subsidy policies, sustainable tourism and

[1]https://orcid.org/0000-0002-3036-0897
[2]https://orcid.org/0000-0003-0512-392

Sustainable Tourism, Part B, 233–250

doi:10.1108/978-1-83608-326-920241014

economic growth. Ultimately, the research aims to provide valuable insights that can guide the formulation of effective subsidy frameworks, fostering a symbiotic relationship between tourism development and long-term environmental and socio-economic sustainability.

Keywords: Sustainability; subsidy; tourism; environment; socio-economic sustainability

1. Introduction

Subsidies emerge as a potential remedy in this complex landscape, offering a mechanism through which governments and stakeholders can actively shape the trajectory of tourism development. The rationale behind subsidies lies in their capacity to influence behaviour, encouraging businesses and tourists to embrace practices that align with sustainability goals. By investigating the effectiveness of subsidy programs, this study seeks to provide empirical insights into whether financial incentives can act as a catalyst for the transition towards a more sustainable tourism paradigm. Gossling and Scott (2012) analyse the role of spatial planning in conjunction with subsidies for sustainable tourism development. The study investigates how subsidies can align with spatial planning strategies to achieve balanced economic growth and environmental conservation in specific regions.

Various case studies, such as the work by Hall and Lew (2018), provide insights into subsidy programs implemented in specific destinations. These case studies offer practical examples of how subsidies have been utilised, their outcomes, and the lessons that can be learnt by policymakers and practitioners. Moreover, the study recognises the multifaceted nature of sustainability in tourism, acknowledging that economic prosperity must be viewed in tandem with social inclusivity and environmental stewardship. Subsidies, therefore, become a tool not only for economic development but also for fostering community resilience, preserving cultural authenticity and protecting natural ecosystems. The examination of subsidy outcomes will shed light on the extent to which these broader sustainability goals are being realised in practice. In their work, Wall and Mathieson (2019) investigate the role of government policies in shaping subsidy programs for sustainable tourism. The study explores how policy frameworks influence the design and effectiveness of subsidies, emphasising the need for a cohesive approach to achieve long-term sustainability goals. Becken and Zhang (2018) investigate how subsidies can be channelised into tourism education and skill development programs. The study assesses the impact of subsidies on enhancing the skill sets of local communities and the tourism workforce, contributing to sustainable development.

Tourism, as a dynamic and multifaceted industry, has emerged as a significant driver of economic growth for nations across the globe. However, the rapid expansion of tourism can pose challenges to environmental preservation, cultural heritage and community well-being. In response to these challenges, the concept of sustainable tourism has gained prominence, emphasising the need for responsible and balanced development. Within the realm of sustainable tourism, the role of subsidies has garnered attention as a potential catalyst for positive change. Subsidies, in the context of tourism, refer to financial incentives, policy frameworks and support mechanisms implemented by governments and other stakeholders to encourage sustainable practices within the industry. Ritchie (2009) explore the role of subsidies in crisis management within the tourism sector. The research investigates how subsidies can be utilised to aid destinations in recovering from crises such as natural disasters, economic downturns or pandemics, while maintaining a focus on sustainability. The overarching goal is to strike a harmonious balance between economic development and the preservation of natural and cultural resources. As we embark on this critical exploration, it is essential to recognise the transformative potential of sustainable tourism. Traditional models of tourism development often led to negative externalities, such as environmental degradation, cultural erosion and strained community relations.

Sustainable tourism, on the other hand, envisions a holistic approach that considers the interconnectedness of economic, social and environmental factors. Subsidies, as instruments of intervention, have been employed by governments and organisations to steer tourism towards sustainability. Financial incentives, tax breaks and regulatory support are among the mechanisms designed to encourage businesses and tourists alike to adopt practices that minimise negative impacts. However, the effectiveness of such subsidies remains an area of debate and warrants an in-depth examination. This study recognises the dynamic nature of the tourism industry and the evolving challenges it faces, including climate change, over-tourism and cultural commodification. Against this backdrop, the role of subsidies takes centre stage as a potential lever for positive change. By critically assessing the outcomes and unintended consequences of subsidy programs, the research aims to contribute to a more nuanced understanding of their role in shaping sustainable tourism paradigms. Furthermore, the study acknowledges the interconnectedness of sustainable tourism with broader economic growth. It seeks to unravel the intricate relationships between subsidy-driven sustainable tourism initiatives and their impact on employment, local economies and the overall well-being of communities. The goal is to identify strategies that not only ensure environmental conservation and cultural preservation but also foster inclusive and equitable economic development. In the pursuit of sustainable tourism, collaboration and engagement among stakeholders – governments, businesses, local communities and tourists – are paramount. This research endeavours to shed light on the collaborative dynamics that influence the success or failure of subsidy programs. By doing so, it aims to provide insights that can guide the formulation of policies and strategies that are not only well-intentioned but also pragmatic and adaptable to diverse context.

The past decades have witnessed an unprecedented surge in global tourism, driven by factors such as increased connectivity, rising disposable incomes, and cultural curiosity. However, this surge has not been without repercussions, as destinations grapple with environmental degradation, loss of biodiversity and challenges to cultural integrity. In addition to the traditional economic indicators, this study also places a spotlight on the social and environmental dimensions of subsidy-driven sustainable tourism. It investigates the extent to which subsidies contribute to community empowerment, equitable distribution of tourism benefits and the reduction of negative environmental externalities. By taking a comprehensive approach, the research aims to provide a holistic understanding of the impacts and implications of subsidy interventions. In conclusion, this critical study on subsidies as a means to promote sustainable tourism for economic growth is positioned at the intersection of environmental conservation, cultural preservation and economic development.

Scheyvens and Biddulph (2018) examine the role of subsidies in supporting community-based tourism initiatives. The study investigates how subsidies can empower local communities economically and socially while preserving their cultural and environmental assets. The significance of this study lies in its exploration of the efficacy of subsidies as a strategic tool for promoting sustainable tourism. As destinations grapple with the dual objectives of fostering economic growth and safeguarding their ecological and cultural assets, understanding the impact of subsidies becomes crucial. The research will contribute to a nuanced understanding of the complexities surrounding subsidy-driven sustainable tourism development, offering insights that can inform policymakers, industry stakeholders and practitioners alike. In navigating the delicate balance between economic prosperity and environmental and cultural conservation, this study seeks to provide a foundation for the refinement and optimisation of subsidy programs. Through a critical lens, the research aims to address gaps in existing subsidy strategies, ultimately guiding the formulation of policies that promote not only immediate economic gains but also enduring benefits for communities, environments and future generations.

Tourism and hospitality is an important parameter of sociocultural identity and heritage of a country. In the era of globalisation tourism and hospitality enhances the economic growth by job creation, source of foreign exchange and development of regions with potential for tourism. According to World Travel and Tourism Council (2014), the contribution of travel and tourism in world GDP is estimated to increase from 9.5% of GDP in 2013 to 10.3% of GDP in 2024. Tourism and hospitality industry contributes 6.8% of GDP of India, contributes 7.7% in total employment generated and provides foreign exchange of US$18.13 billion. The tourism and hospitality sector is the third largest source of foreign exchange for India. The integration of circular economy principles in tourism through subsidies. The study examines how subsidies can incentivise businesses to adopt circular practices, reducing waste and enhancing sustainability in the tourism supply chain. Saarinen and Rogerson (2018) investigate how subsidies can contribute to inclusive tourism development. The study explores ways in which subsidies can address social equity concerns, ensuring that the benefits of tourism development are shared by diverse segments of the population.

2. Review of Literature

A comprehensive literature review on the topic of 'Subsidy as a Means to Promote Sustainable Tourism for the Development of Economic Growth' provides an in-depth understanding of the existing research, theories and findings related to this subject. The review covers various aspects, including the role of subsidies in sustainable tourism, their impact on economic development and the challenges associated with their implementation. Weaver and Lück (2018). Within the intricate tapestry of global tourism, the quest for sustainability is evolving into a pressing imperative. Traditional tourism models have often been characterised by a linear progression, prioritising sheer numbers of visitors and economic gains without due consideration for the environmental and sociocultural repercussions. This study recognises the paradigm shift towards sustainable tourism as an intricate dance, requiring careful choreography between economic development, ecological preservation and social equity. Subsidies, as instruments of change, play a pivotal role in orchestrating this delicate balance. They operate as the silent architects, shaping the contours of sustainable tourism by incentivising responsible practices among businesses and travelers. As we embark on this critical exploration, the study seeks to unravel the layers of subsidy mechanisms, from financial incentives to policy frameworks, and assess their effectiveness in instigating positive change within the tourism landscape. A study by Gössling and Peeters (2015) explores the various types of subsidies employed in the context of sustainable tourism, emphasising financial incentives, tax breaks and regulatory support. The research delves into the motivations behind governments' use of subsidies and the mechanisms through which they influence sustainable practices. Dwyer and Forsyth (2018) investigate the economic implications of subsidies within the tourism sector. The study assesses the direct and indirect effects of subsidies on job creation, income generation and overall economic growth, offering insights into the potential contributions of subsidy programs to destination economies. Research by Jamal and Stronza (2016) explores the social and cultural aspects of sustainable tourism, specifically addressing how subsidies can be designed to benefit local communities. The study investigates the role of subsidies in preserving cultural heritage, empowering communities and fostering positive social impacts. Gössling et al. (2017) examine the importance of stakeholder collaboration and effective governance in subsidy-driven sustainable tourism. The research emphasises the need for coordination among governments, businesses, local communities and tourists to maximize the positive impacts of subsidies. By synthesising these studies, the literature review contributes to a holistic understanding of the complexities surrounding subsidies as a means to promote sustainable tourism for economic growth. It provides a foundation for identifying gaps in current knowledge, framing research questions and guiding the direction of future studies in this critical field. Gössling and Scott (2020) delve into the behavioral aspects of tourists in response to subsidies promoting sustainable practices. This research examines how tourists perceive and respond to subsidised initiatives, shedding light on the potential for behavioral change and the factors influencing tourist decision-making. Matarrita-Cascante and Brennan (2017) provide insights into the methodologies and tools used to measure the impact of subsidies on sustainable tourism. The study addresses challenges in evaluating the success of subsidy programs and proposes frameworks for assessing their effectiveness in achieving economic and sustainability objectives. A global perspective is offered by Weaver and Lück (2018), who analyze subsidy programs across diverse international contexts. The research identifies commonalities and differences in subsidy strategies, drawing attention to the role of cultural, economic and environmental factors in shaping subsidy outcomes on a global scale. Buhalis and Leung (2018) explore the integration of technological innovations in subsidy-driven sustainable tourism. The study investigates how advancements such as digital platforms and data analytics can enhance the efficiency and impact of subsidy programs, fostering a technologically enabled approach to sustainability. Becken and Hughey (2020) investigate the long-term sustainability and resilience of destinations under subsidy-driven sustainable tourism. The research assesses the enduring impacts of subsidy programs, considering factors such as adaptive capacity, community resilience and the ability of destinations to cope with external shocks. Gössling and Higham (2019) analyze the role of public–private partnerships in the implementation of subsidy programs. The study highlights the collaborative efforts between governments and private enterprises, emphasising the importance of joint initiatives in achieving sustainable tourism objectives. The governance structures that influence subsidy programs in tourism. The research assesses the role of effective governance in shaping subsidy outcomes, emphasising the need for transparent and inclusive decision-making processes involving multiple stakeholders. Scheyvens and Biddulph (2018) examine the role of subsidies in supporting community-based tourism initiatives. The study investigates how subsidies can empower local communities economically and socially while preserving their cultural and environmental assets. Ritchie (2009) explore the role of subsidies in crisis management within the tourism sector. The research investigates how subsidies can be utilised to aid destinations in recovering from crises such as natural disasters, economic downturns or pandemics, while maintaining a focus on sustainability. Becken and Zhang (2018) investigate how subsidies can be channeled into tourism education and skill development programs. The study assesses the impact of subsidies on enhancing the skill sets of local communities and the tourism workforce, contributing to sustainable development. Whitford et al. (2017) focus on indigenous tourism and analyze the effectiveness of subsidies in preserving indigenous cultures while fostering economic development. The research explores the challenges and opportunities of implementing subsidy programs in the context of indigenous tourism. Ioannides and Petersen (2018) explore the relationship between subsidies, innovation and entrepreneurship in the tourism industry. The study investigates how subsidies can encourage the adoption of innovative practices and entrepreneurial initiatives, leading to sustainable tourism development. The role of subsidies in destination marketing for sustainable tourism. The study analyzes how subsidies can be strategically utilised to promote destinations that prioritise environmental responsibility, cultural preservation and community engagement. Miller et al. (2017) assess the impact of subsidies on encouraging tourism businesses to attain responsible tourism certifications. The study explores the relationship between subsidies and certification programs as tools for promoting sustainable practices within the industry. Kim and Hall (2018) investigate the impact of subsidised sustainability initiatives on tourist satisfaction. The research explores how tourists' perceptions of destinations are influenced by subsidised programs that emphasise environmental conservation, community engagement and cultural preservation. Gössling et al. (2021) explore the integration of circular economy principles in tourism through subsidies. The study examines how subsidies can incentivise businesses to adopt circular practices, reducing waste and enhancing sustainability in the tourism supply chain. Scott et al. (2019) analyse the role of spatial planning in conjunction with subsidies for sustainable tourism development. The study investigates how subsidies can align with spatial planning strategies to achieve balanced economic growth and environmental conservation in specific regions. Saarinen and Rogerson (2018) investigate how subsidies can contribute to inclusive tourism development. The study explores ways in which subsidies can address social equity concerns, ensuring that the benefits of tourism development are shared by diverse segments of the population. Hall and Amelung (2018) examine the role of subsidies in promoting sustainable transportation options for tourists. The research assesses how subsidies can incentivise the use of eco-friendly transportation modes, contributing to the reduction of the carbon footprint in tourist travel. Gössling and Michael Hall (2019) explore the challenges and opportunities of implementing subsidies in the context of cross-border tourism. The study analyses how collaboration between nations, facilitated by subsidies, can lead to sustainable tourism development in shared destinations. Neuhofer et al. (2015) investigate the integration of technology in subsidised tourism experiences. The study explores how technological innovations, when coupled with subsidies, can enhance the overall quality and sustainability of the tourist experience. Butler and Hinch (2017) examine the impacts of subsidies on niche tourism markets. The research explores how subsidies can support the development of niche tourism sectors, fostering unique and sustainable offerings that cater to specialised interests.

3. Objectives of the Study

- To assess the awareness level of tribal people about government sponsored subsidies.
- To examine the impact of subsidy schemes on economic development of tribal communities semiliguda block of Koraput district.

4. Hypothesis

H1. There is no significant association between awareness levels of tribal people with tourism subsidies awareness.

H2. There will be a significant association between awareness levels of income with tourism subsidies awareness.

5. Sources of Data Collection

Primary data is the main source for this study and it will be collected through questionnaire method. Data will be collected from tribal people from three villages of Semiliguda block and through direct interaction with respondents of three villages. Secondary data will be collected from different sources like journals, newspapers, magazines, literature reviews from various authors and internet websites.

6. Methodology

A field survey has been conducted in Koraput district of Odisha. Research data will be collected from one block out of which the Researcher has taken three villages for this study. This block represents different levels of subsidies on tourism given by state and central government particularly to tribal people of Semiliguda block. To achieve the objective of research study the researcher will use field survey method and questionnaire method. And further, data will be collected from 80 respondent respondents by asking subsidy related questions. See Table 14.1.

Table 14.1. Educational Qualification of Respondents in Semiliguda Block.

Educational Qualification of Respondent Respondents	Number of Respondents	Respondents in %
Uneducated	12	15%
Primary education	25	31%
Secondary education	37	46%
Graduation	06	08%
Postgraduation	00	00%
Other	00	00%
Total	80	100%

Source: Data collected from the field sources.

7. Data Analysis

As Fig. 14.1 shows, we found that highest 46% of respondents belong to secondary education, 15% of the respondents belong to uneducated, 31% of respondents belong to primary education and 8% of the respondents belong to graduation level. Table 14.2 shows the annual income level of respondents.

From Fig. 14.2, we found that highest 65% of respondents in Semiliguda block have monthly income of 50,000 or less than 50,000, 23% of respondents have monthly income of 50,000–80,000, 10% of respondents have monthly income of 80,000–1,20,000 and 2% of respondents have monthly income of more than 1,20,000. See Table 14.3 to see types of subsidies on tourism adopted by respondents in Koraput district.

From Fig. 14.3, we found that highest 73% of respondents depend on Hotel and Accommodation Subsidies, 23% of respondents are depending on Infrastructure Development and 4% of the respondents depend on Tourism Marketing and Promotion. See Table 14.4 for the accommodations of tourists in Semiliguda block.

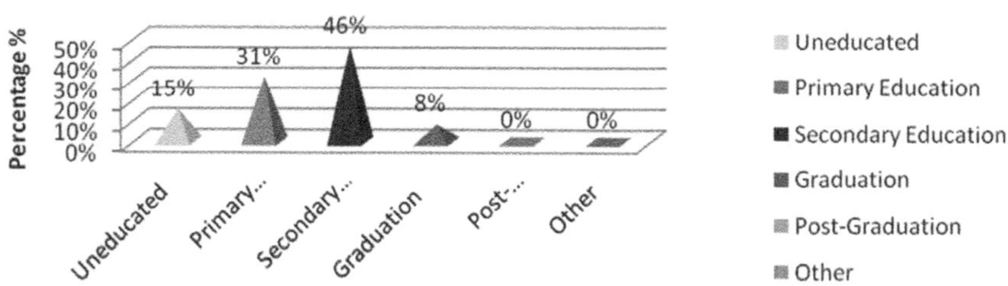

Fig. 14.1. Educational Qualification of Respondents in Semiliguda Block.

Table 14.2. Annual Income of the Respondent Respondents of Semiliguda Block.

Annual Income of Respondents	Number of Respondents	Number of Respondents in %
≤50,000	52	65%
50,000–80,000	18	23%
80,000–1,20,000	08	10%
Above 1,20,000	2	2%
Total	80	100%

Source: Data collected from the field sources.

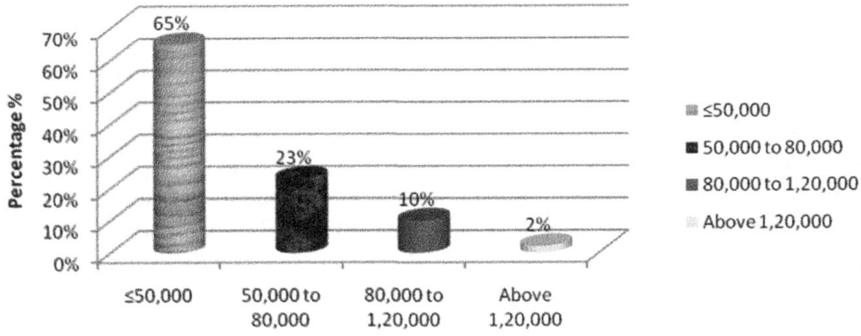

Fig. 14.2. Annual Income of the Respondents of Semiliguda Block.

Table 14.3. Types of Subsidies on Tourism Adopted by Respondents in Koraput District.

Types of Subsidies	**Number of Respondents**	**Number of Respondents in %**
Infrastructure development	19	23%
Hotel and accommodation subsidies	58	73%
Tourism marketing and promotion	3	4%
Environmental and cultural conservation	00	00%
Total	80	100%

Source: Data collected from the field sources.

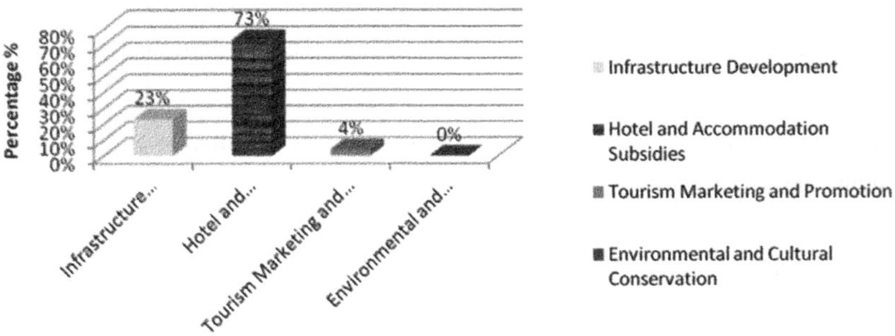

Fig. 14.3. Types of Subsidies on Tourism Adopted by Respondents in Koraput District.

Table 14.4. Accommodations of Tourists in Semiliguda Block.

Accommodations of Tourists	Number of Respondents	Number of Respondents in %
Quality of service	37	46%
Price	32	40%
Safety and security	11	14%
Reviews and ratings	00	00
Total	80	100%

Source: Data collected from the field sources.

Fig. 14.4 reveals that highest 46% of respondents satisfied on Quality of Service, 40% of respondents satisfied on pricing for accommodation and 14% of respondents have faith on Safety and Security. Table 14.5 shows the awareness about subsidies on tourism of Semiliguda block.

Fig. 14.5 reveals that 94% of the respondents are not aware of the subsidies on tourism system and 6% of the respondents are aware about the subsidies on tourism, while Table 14.6 shows subsidy received by respondents in Semiliguda block.

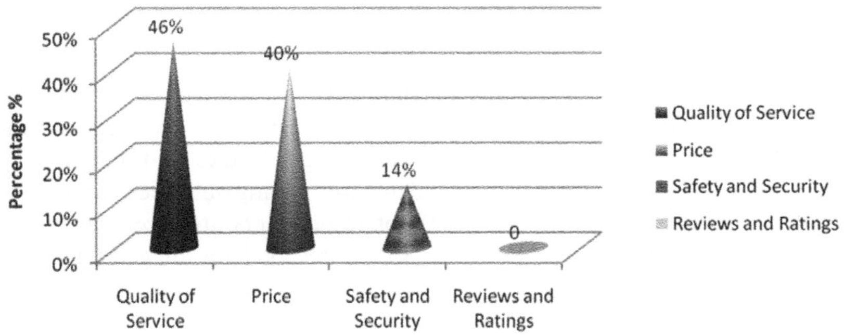

Fig. 14.4. Accommodations of Tourists in Semiliguda Block.

Table 14.5. Awareness About Subsidies on Tourism of Semiliguda Block.

Awareness About Subsidies on Tourism	Number of Respondents	Number of Respondents in %
Yes	05	6%
No	75	94%
Total	80	100%

Source: Data collected from the field sources.

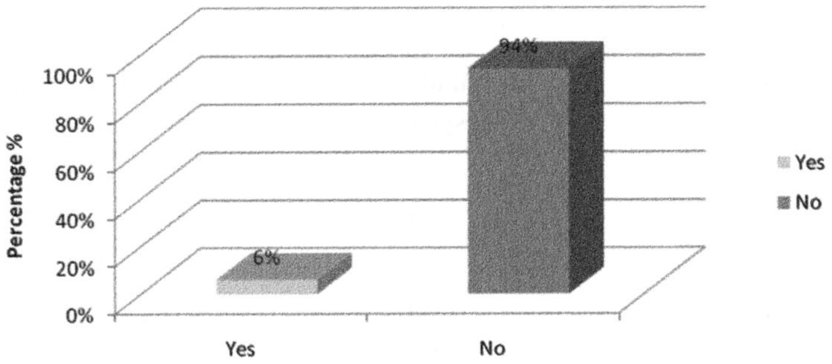

Fig. 14.5. Awareness About Subsidies on Tourism of Semiliguda Block.

Table 14.6. Subsidy Received by Respondents in Semiliguda Block.

Subsidy Received by Respondents	Number of Respondents	Number of Respondents in %
Yes	08	12%
No	72	88%
Total	80	100%

Source: Data collected from the field sources.

Fig. 14.6 depicts that 88% of respondents have not received any kind of subsidy from the government and 12% of the respondents received subsidies. Table 14.7 shows types of agencies for distributing subsidies on tourism.

Fig. 14.7 depicts that 61% of respondents are getting subsidies from Collaboration with Industry Stakeholders, 19% of respondents are receiving subsidies from Government Policies and Priorities, 15% of respondents are getting subsidies

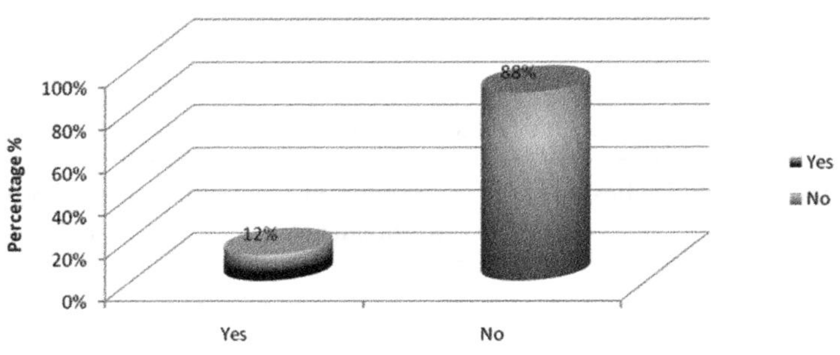

Fig. 14.6. Subsidy Received by Respondents in Semiliguda Block.

Table 14.7. Types of Agencies for Distributing Subsidies on Tourism.

Different Distributing Agencies	Number of Respondents	Number of Respondents in %
Government policies and priorities	15	19%
Sustainability initiatives	03	5%
Crisis response	13	15%
Collaboration with industry stakeholders	49	61%
Total	80	100%

Source: Data collected from the field sources.

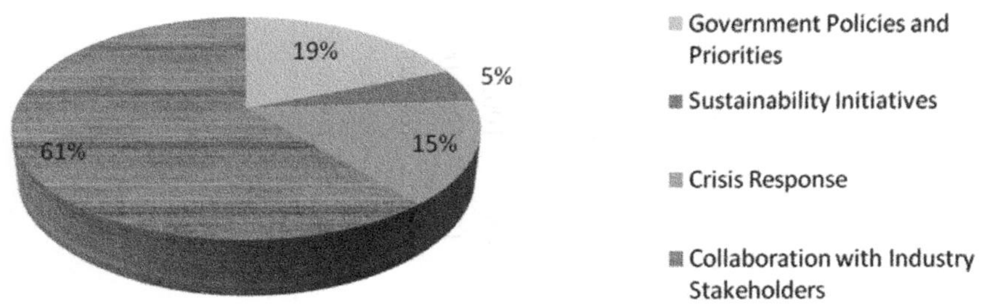

Fig. 14.7. Different Distributing Agencies for Distributing Subsidies on Tourism.

from Crisis Response and 5% of respondents are getting subsidies from Sustainability Initiatives.

Table 14.8 shows the association between the levels of income with standard of living. We found from the Chi-square table p -value is 0.331 which is greater than 0.05. So it is not significant. But the null hypothesis is accepted that means there will be no significant association between the levels of income and standard of living. With respect to level of income there is a no significant difference among their perception regarding the impact of subsidies on tourism and the standard of living. Most of the respondents, across different income groups, think in the same way that subsidy will increase the standard of living of respondents.

Table 14.9 shows the association between the level of income of respondents with level of awareness. We found from the Chi-square table p -value is 0.556 which is greater than 0.05. So it is not significant. But the null hypothesis is accepted that means there will be no significant association between the level of income of respondents and level of awareness.

Table 14.8. Association Between Level of Income and Their Standard of Living.

Standard of Living									
	No		**Yes**		**Total**		**Chi-Square Value**	**Df**	***p* Value**
Levels of Income	**F**	**%**	**F**	**%**					
≤50,000	15	31.25%	33	68.75%	48	100.00%	3.425	3	0.331
50,000–80,000	7	41.17%	10	58.83%	17	100.00%			
80,000–1,20,000	09	64.28%	5	35.72%	14	100.00%			
Above 1,20,000	0	0.00%	1	100.00%	1	100.00%			
Total	31		49		80	100.00%			

Source: Data collected from the field sources.

Table 14.9. Association Between Respondents Level of Income With Awareness Level About Subsidies on Tourism.

Awareness Level	No		Yes		Total		Chi-Square Value	Df	p Value
Level of Income	F	%	F	%					
≤50,000	5	26.31%	14	73.69%	19	100.00%	6.825	8	0.556
50,000–80,000	7	35%	13	65%	20	100.00%			
80,000–1,20,000	8	38.09%	13	61.91%	21	100.00%			
Above 1,20,000	1	11.11%	8	88.89%	09	100.00%			
Above 1,20,000	1	9.09%	10	90.91%	11	100.00%			

Note: $p \leq 0.05$ (significant).

8. Conclusion

The researcher has found that in Semiliguda block a very smaller number of respondents are receiving agricultural subsidy. Respondents mainly received subsidies from stakeholders and Government agencies with the help of crises management. But the respondents of Semiliguda block are not satisfied with distributing system. So, lack of awareness is major hurdle faced by respondents for getting subsidies. Respondents are not getting the whole benefit of subsidy due to some loopholes and corrupt practices in the system. Respondents in Semiliguda block, have reported are not receiving the allied businesses including poultry, that they were suppose to get. Subsidies in tourism not only reduce suicides among tribal communities but also foster their growth in Semliguda block through initiatives like Infrastructure development, hotel and accommodation subsidies, tourism marketing and promotion and environmental and cultural conservation efforts.

Bibliography

Becken, S., & Hughey, K. F. (2016). *Tourism and water*. Channel View Publications.

Becken, S., & Hughey, K. F. (2020). Tourism and water: Interactions and impacts—An introduction. *Journal of Sustainable Tourism*, *28*(5), 601–617.

Becken, S., & Zhang, G. (2018). Progress in tourism management: A review of innovation research in tourism. *Tourism Management*, *68*, 52–73.

Buhalis, D., & Leung, D. (2018). Smart hospitality—Interconnectivity and interoperability towards an ecosystem. *International Journal of Hospitality Management*, *71*, 41–50.

Butler, R., & Hinch, T. (2017). *Tourism and indigenous peoples: Issues and implications*. Routledge.

Dredge, D., & Gyimóthy, S. (2015). *Tourism collaboration and partnerships: Politics, practice and sustainability*. Channel View Publications.

Dwyer, L., & Forsyth, P. (2018). The economic impacts of tourism subsidies: A computable general equilibrium analysis. *Journal of Travel Research*, *57*(5), 631–646.

Gössling, S., & Peeters, P. (2015). Subsidies for environmentally sustainable tourism: A review. *Tourism Management*, *46*, 386–398.

Gössling, S., & Higham, J. (2019). *Tourism and regional development: New pathways*. Routledge.

Gössling, S., & Scott, D. (2012). *Tourism and water: Interactions and impacts*. Channel View Publications.

Gössling, S., & Scott, D. (2020). Tourism and water: Interactions and impacts – An introduction. *Journal of Sustainable Tourism*, *28*(5), 601–617.

Gössling, S., Scott, D., & Hall, C. M. (2012). *Tourism and water*. Channel View Publications.

Gössling, S., Scott, D., & Hall, C. M. (2013). *Tourism and water: Interactions and impacts*. Channel View Publications.

Gössling, S., & Michael Hall, C. (2019). Sharing versus collaborative economy: how to align ICT developments and the SDGs in tourism? *Journal of Sustainable Tourism*, *27*(1), 74–96.

Hall, C. M., & Amelung, B. (2018). *Tourism and regional development: New pathways*. Routledge.

Hall, C. M., & Lew, A. A. (2018). *Sustainable tourism: A marketing perspective*. Routledge.

Higham, J., & Lück, M. (2019). *Marine wildlife tourism*. CABI.

Ioannides, D., & Gyimóthy, S. (2020). The COVID-19 crisis as an opportunity for escaping the unsustainable global tourism path. *Tourism Geographies*, *22*(3), 624–632.

Ioannides, D., & Petersen, L. (2018). Reframing tourism innovation: A response to post-2010 tourism development. *Current Issues in Tourism*, *21*(15), 1739–1760.

Jamal, T. B., & Stronza, A. L. (2012). Collaboration theory and tourism practice in protected areas: Stakeholders, structuring and subsidies. *Current Issues in Tourism*, *15*(3), 211–220.

Jamal, T., & Stronza, A. (2016). Collaboration theory and tourism practice in protected areas: Stakeholders, structuring and subsidies. *Current Issues in Tourism*, *19*(6), 531–551.

Jamal, T., & Camargo, B. A. (2019). *Tourism and development in the global south: The issues*. Routledge.

Kim, M. J., & Hall, C. M. (2018). A hedonic motivation model in virtual reality tourism: Comparing visitors and non-visitors. *International Journal of Information Management*, *46*, 236–249.

Matarrita-Cascante, D., Brennan, M. A., & Luloff, A. E. (2015). Community-based tourism development: A conceptual framework for understanding collaboration with rural communities. *Journal of Sustainable Tourism*, *23*(6), 869–889.

Matarrita-Cascante, D., & Brennan, M. A. (2017). Towards understanding the dynamics of community-based tourism development: A 25-year review. *Journal of Sustainable Tourism*, *25*(7), 869–890.

Miller, M. L., et al. (2017). *Tourism and regional development: New pathways*. Routledge.

Neuhofer, B. Buhalis, D., & Ladkin, A. (2015). Technology as a catalyst of change: Enablers and barriers of the tourist experience and their consequences. In *Information and Communication Technologies in Tourism 2015: Proceedings of the International Conference in Lugano*, Switzerland, February 3–6, (pp. 789–802). Springer International Publishing.

Ritchie, B. W. (2009). *Crisis and disaster management for tourism*. Multilingual Matters.

Saarinen, J., & Rogerson, C. M. (2018). *Rural tourism development: Localism and cultural change*. Routledge.

Scheyvens, R., & Biddulph, R. (2018). Inclusive tourism development. *Tourism Geographies*, *20*(4), 589–609.

Scott, D., & Gössling, S. (2021). Tourism and water: Interactions and impacts—An introduction. *Journal of Sustainable Tourism*, *29*(1), 1–19.

Wall, G., & Mathieson, A. (2019). *Tourism: Change, impacts and opportunities*. Pearson.

Weaver, D. B., & Lück, M. (2017). The dimensions of sustainable tourism. In *The Routledge handbook of transport economics* (pp. 348–364). Routledge.

Weaver, D. B., & Lück, M. (2018). *The politics of tourism in Asia*. Channel View Publications.

Whitford, M., Ruhanen, L., & Carr, A. (Eds.). (2017). *Indigenous tourism: Cases from Australia and New Zealand*. Goodfellow Publishers Ltd.

Chapter 15

Catalysing Destination Sustainability: Role of Self-Help Groups

Shreeansh Mishra[a,1], *Jitendra Mohan Mishra*[b,2] *and Vaibhav Bhatt*[a,3]

[a]Central University of Tamil Nadu, India
[b]Central Tribal University of Andhra Pradesh, India

Abstract

This research article explores the intricate relationship between self-help groups (SHGs) and destination sustainability in the context of tourism. SHGs, typically formed by individuals with shared interests and objectives, have gained prominence as a means to promote economic, social and environmental sustainability at tourism destinations. This study investigates the impact of SHGs on destination sustainability, focusing on various dimensions, including economic empowerment, cultural preservation, environmental conservation and social development. After conducting an extensive review of the literature and supported by empirical case studies, this chapter undertakes a comprehensive analysis of the role played by SHGs in promoting the sustainability of tourism destinations. Furthermore, it examines the challenges encountered by SHGs in their pursuit of sustainable tourism objectives. The findings of this research contribute to a better understanding of the potential of SHGs in fostering destination sustainability and provide valuable insights for policymakers, destination managers and stakeholders interested in harnessing the power of community-driven initiatives for sustainable tourism.

Keywords: Self-help groups; destination sustainability; community empowerment; sustainable tourism; social development; economic impact; cultural preservation; environmental conservation

[1]https://orcid.org/0009-0004-8847-6587
[2]https://orcid.org/0000-0002-5840-9701
[3]https://orcid.org/0000-0003-2859-7913

Sustainable Tourism, Part B, 251–263

Published under exclusive licence by Emerald Publishing Limited
doi:10.1108/978-1-83608-326-920241015

1. Introduction

In an era characterised by rapid globalisation and increased human mobility, the sustainable development of tourism destinations has become an imperative concern. The intersection of community-driven initiatives and sustainable tourism has emerged as a focal point for scholars, policymakers and practitioners seeking innovative approaches to address the complex challenges inherent in contemporary tourism development (Lim & Lee, 2020). Destination sustainability emerged as a specific application of sustainable development principles (Nadalipour et al., 2019; Than et al., 2020), focusing on creating and maintaining tourism practices that enhance the well-being of local communities, preserve cultural heritage, protect the environment and provide a positive and fulfiling experience for visitors (Mathew & Sreejesh, 2017; Than et al., 2020; Venugopalan, 2021). Destination sustainability is a holistic and evolving concept that encapsulates the responsible management of tourism activities to ensure the long-term well-being of both the environment and local communities (Artal-Tur et al., 2018; Carey et al., 1997). Striking a delicate balance between economic development and environmental stewardship, destination sustainability seeks to minimise negative impacts such as over-tourism, environmental degradation and cultural erosion (Ibnou-Laaroussi et al., 2020; Nanda Praptiwi et al., 2022). It involves fostering a symbiotic relationship between tourists and host communities, promoting ethical tourism practices and embracing innovative solutions to enhance the resilience and longevity of destinations for future generations.

The traditional top-down approach, which emphasised the role of experts and government agencies, has evolved into an understanding that thriving local communities play a crucial role in sustainable tourism development (Than et al., 2020). Furthermore, the empowerment of community groups around tourist destinations is vital in enhancing tourist attractions (Simpson, 2001; Lee, 2013). These community-led initiatives contribute to maintaining a positive environment, improving the economic conditions of the community and preserving the region's traditions and cultures. One such initiative is self-help groups (SHGs), which have played a pivotal role in empowering communities and facilitating their participation in sustainable tourism development.

SHGs gained popularity in developing countries as one of the key mechanisms for community empowerment and participation in sustainable tourism development (Vinodan & Meera, 2022). These groups are typically formed by individuals from the local community who come together to address common issues and work towards their socio-economic development (Bwala et al., 2016; Chitale, 2017). Through collective efforts and mutual support, SHGs facilitate the appropriation of resources and mobilisation of locally centric knowledge, thus catalysing destination sustainability within the context of tourism (Vinodan & Meera, 2022). SHGs operate on the principles of collective action, mutual support and shared decision-making.

Several empirical studies have emphasised the importance of local and community participation in the responsible management of tourism sites for sustainable tourism development (Stefanica & Vlavian-Gurmeza, 2010; Wei et al., 2021). In this direction, the study tried to explore the role of SHGs as grassroots-level institutions for catalysing the efforts to achieve destination sustainability. This study is significant as it helps in understanding a community-centric approach directed towards destination sustainability.

2. Literature Review

The literature on the role of SHGs in sustainable development has grown significantly in recent years (Gaas, 2019; Kalra et al., 2013; Kumar et al., 2021), reflecting a growing recognition of the potential impact of community-driven initiatives on the tourism sector (Freitas & Koskowski, 2021). Several scholars emphasise the pivotal role of SHGs in empowering local communities(Bwala et al., 2016; Gaas, 2019) and fostering active participation in sustainable tourism initiatives (Mishra & Mishra, 2023). These studies highlight SHGs as effective platforms for community engagement enabling members to actively contribute to decision-making processes, resource management and the preservation of cultural heritage. There are studies emphasising the contribution of SHGs to the economic upliftment of local communities (Jayant & Dev Bhatt, 2021; Samal & Mohopatra, 2016). Income-generating activities, such as handicraft production, homestays and guided tours facilitated by SHGs, are shown to provide alternative livelihood options and contribute to poverty alleviation (Ravindran & Vinodan, 2009). Moreover, SHGs have been recognised as key actors in promoting responsible tourism practices and environmental conservation efforts (Vinodan & Meera, 2022). Training in hospitality, entrepreneurship and sustainable practices by SHGs has been identified as instrumental in enhancing the quality of services and positioning local communities as active stakeholders in tourism development (Vukovic et al., 2021). Nichols (2021), in his study, highlights the contribution of SHGs to social capital within communities. Collaborative efforts and shared responsibilities within SHGs are shown to strengthen community bonds, leading to improved social well-being and a sense of collective responsibility (Bwala et al., 2016; Ruducha et al., 2019).

It has been noted that while there are numerous individual studies on SHGs as a means of economic and social development, there is currently no comprehensive study that emphasises their contribution to the sustainable development of destinations. The existing literature often lacks an in-depth exploration of various ways in which SHGs support sustainable tourism development. Therefore, further academic investigation is essential to elucidate the ways in which the operations and outcomes of SHGs impact the communities and societies they engage within the realm of tourism. Consequently, it is imperative to identify the challenges hindering SHGs in their progressive efforts to engage and catalyze destination sustainability.

3. Research Objective

The primary objective of this study is to systematically identify and analyse the intricate role of SHGs in contributing to destination sustainability. Accordingly, this study discerns the diverse challenges encountered by SHGs in their involvement in the sustainable development of tourism destinations.

4. Methodology

To achieve the research objectives, this study will adopt a qualitative approach. This approach will include a literature review and observations from the case studies. The study will begin with a comprehensive review of existing literature on the role of SHGs in sustainable tourism development. The literature review will provide a theoretical foundation for understanding the linkages between SHGs and sustainable tourism development in tourism destinations. It will also identify key concepts, frameworks and theories that can guide the empirical investigation. Following the literature review, contextual case studies are studied and observations are made to answer the research objectives.

5. Findings and Discussions

An in-depth exploration of contextual literature and case studies from India reveals the diverse contributions of SHGs in stimulating destination sustainability within the tourism context. Four key elements necessary to accomplish sustainable tourism development in destination include economic empowerment, preservation of culture, conservation of the environment and social progress (Pan et al., 2018) as shown in Fig. 15.1. The findings suggest that SHGs play a crucial role in economic empowerment, cultural preservation, environmental conservation and social development, contributing to sustainable tourism development in destinations.

6. Destination Sustainability and Self-Help Groups

Economic Empowerment: One of the key contributions of SHGs to sustainable tourism development is economic empowerment. SHGs contribute to economic empowerment by creating employment opportunities and generating income for local communities (Chitale, 2017; Jayant & Dev Bhatt, 2021). These groups offer opportunities for residents to engage in tourism-related activities and generate income (Ravindran & Vinodan, 2009). Furthermore, SHGs promote entrepreneurship and provide access to credit facilities, training programs and market linkages for community members (Ahemad & Katoch, 2022). For example, SHGs in tourism destinations often engage in entrepreneurial activities such as running homestays, local handicraft production and providing tour guide services (as mentioned in case study 1). This economic empowerment not only enhances the livelihoods of individuals but also contributes to poverty eradication and economic growth at the community level. Furthermore, SHGs promote entrepreneurship and provide access to credit facilities, training programs and market linkages for community members (Ahemad & Katoch, 2022; Kumar et al., 2021), which is crucial for fostering economic empowerment and community development.

Fig. 15.1. Four Key Elements Necessary to Accomplish Sustainable Tourism Development. *Source:* Adapted from: Pan et al. (2018).

Cultural Preservation: Another significant contribution of SHGs to sustainable tourism development is the preservation and promotion of local culture. SHGs play a crucial role in preserving traditional knowledge, skills and practices that are integral to the cultural identity of a community (Vinodan & Meera, 2022). By actively involving members of the community in tourism development, SHGs ensure that the local culture is showcased and celebrated (Vukovic et al., 2021). This not only helps in the preservation of cultural heritage but also creates opportunities for cultural exchange and understanding among tourists and locals. The involvement of SHGs in cultural preservation can be seen through activities such as organising cultural events, promoting traditional arts and crafts and offering cultural experiences such as folklore performances, traditional dance and music and culinary tours. Case Study 2 underscores the intervention of SHGs in preserving and enhancing artisanal skills, thereby contributing to the maintenance of cultural sustainability. Similarly, a case study conducted in a heritage site in Kerala, India, found that SHGs were actively involved in the development and promotion of local art forms and indigenous products through their participation in exhibitions and business meets (Vinodan & Meera, 2022).

Environmental Conservation: SHGs make valuable contributions to the conservation of the environment and natural resources. They have long been instrumental in environmental conservation and sustainable practices, particularly in tourism destinations, and are often engaged in activities such as cleanliness drives around tourism destinations, waste management and promoting sustainable practices among community members and tourism stakeholders (Kandpal & Saizen, 2022; Vukovic et al., 2021). By raising awareness and implementing environmentally friendly practices, SHGs contribute to the preservation of natural resources and the reduction of negative impacts on the environment as exemplified in Case Study 1. Similarly, in ecotourism destinations, SHGs conduct sessions focused on complying with national/state environment protection norms and promoting sustainable waste management practices such as recycling and composting. Their participation in exhibitions and business meets has further solidified their role in advocating for eco-friendly measures. By actively engaging in activities such as cleanliness drives and waste management, SHGs have made significant contributions to preserving natural resources. These efforts have not only reduced the negative impacts on the environment but have also led to increased awareness among community members and tourism stakeholders.

Social Development: SHGs play a crucial role in empowering marginalised communities and promoting social development. SHGs provide a platform for community members to come together, share knowledge and resources and address their common social issues (Bwala et al., 2016; Ruducha et al., 2019) This leads to enhanced social cohesion, increased self-confidence, social capital and improved decision-making abilities among the members (Kumar & Jasheena, 2016; Nichols, 2021). Furthermore, SHGs often provide skill development training and capacity-building programs to their members, which not only enhances their employability but also promotes entrepreneurship and self-employment opportunities (Gaas, 2019; Kumar & Jasheena, 2016; Samishetti & Rao, 2018) Moreover SHGs have indeed played a pivotal role in addressing the challenges faced by women in various communities, contributing to their empowerment and socio-economic development (Ahemad & Katoch, 2022; Kumari, 2014; Maina, 2021; Ruducha et al., 2019). Through these initiatives, SHGs contribute to poverty alleviation, gender equality and overall social well-being in tourism destinations. Both case studies vividly depicted how SHGs catalyze social development by empowering communities and bolstering social capital.

Case Study 1: Swamini Mangrove Safari – Fostering Sustainable Tourism in Sindhudurg, Maharashtra, India

Situated in the southernmost district of Maharashtra, Sindhudurg, despite covering only 3.8% of the state's total mangrove vegetation, stands out as the richest in biodiversity. The transformative journey of the 'Swamini' SHG, led by Shweta Hule, and their pioneering initiative – the 'Swamini Mangrove Safari' in the Mandavi Creek of Vengurla taluka is a perfect example of sustainable tourism development mobilising community support. Since 2017, Swamini has been organising mangrove safaris, recognised as a model for community-led conservation through ecotourism. Recognising their success, the State Forest Department has sought to replicate this model across coastal Maharashtra. The safari, operating with support from the Mangrove Cell of Maharashtra and UNDP India, involves rowboat rides guided by Swamini members, offering tourists an immersive experience in the rich biodiversity of the Mandavi Creek. Notably, the Swamini Group's members, initially with limited knowledge of mangroves, underwent a remarkable transformation. Encouraged by the Mangrove Cell and UNDP-GEF Project, they not only learnt to row boats but also identified the eight species of mangroves in the creek by their local and scientific names. This initiative not only serves as a successful ecotourism model but also highlights the potential of SHGs in driving sustainable tourism development, emphasising community engagement and environmental conservation.

Source: (Mongabay, 2020).

Case Study 2: DHAN Foundation's Tourism for Development Initiative

DHAN Foundation's 'Tourism for Development' initiative stems from a successful pilot project in Tamil Nadu and seeks to harness the potential of the tourism sector for poverty reduction. The initiative employs a multifaceted approach, including organising poor families into tourism-based People's Organisations, conserving heritage and artisanal skills, enhancing skills related to tourism, linking communities with external stakeholders, supporting research and development and promoting responsible tourism practices. Through these strategies, DHAN Foundation aims to empower local communities economically, preserve cultural identity, develop new skills, establish market linkages and foster sustainable tourism practices.

This comprehensive approach has yielded positive outcomes, including economic empowerment, cultural conservation, skill development, enhanced market access and the integration of responsible tourism practices. By emphasising the role of SHGs, DHAN Foundation's initiative serves as a model for how community-driven efforts can contribute significantly to sustainable tourism development, creating a holistic and inclusive impact on both local communities and the tourism sector.

Source: (DHAN Foundation, 2023).

The findings underscore the pioneering role of SHGs in advancing destination sustainability through multifaceted contributions within the framework of sustainable tourism development. Notably, SHGs significantly contribute to economic empowerment by creating employment opportunities, fostering entrepreneurship and providing essential support systems for community development. They emerge as key custodians of cultural heritage, actively preserving and promoting local traditions and practices, thereby fostering cultural exchange and understanding between tourists and local communities. Additionally, the research highlights the innovative environmental conservation efforts of SHGs, advocating for sustainable practices and actively participating in activities such as cleanliness drives and waste management. Furthermore, SHGs play a crucial role in social development by empowering marginalised communities, enhancing social cohesion and addressing common social issues. Their distinct impact on women's empowerment and socio-economic development contributes significantly to poverty alleviation and overall social well-being in tourism destinations. In essence, the study unveils the transformative and holistic role of SHGs in catalysing destination sustainability across economic, cultural, environmental and social dimensions within the context of sustainable tourism development.

7. Challenges Encountered

While SHGs have made significant contributions to sustainable tourism development, they also face several challenges that hinder their full potential. The challenges encountered by SHGs in the pursuit of sustainable tourism development are indeed multifaceted, necessitating a profound comprehension of the local context and industry dynamics. These challenges include:

- *Financial Constraints:* Limited access to funding and financial resources can impede the ability of SHGs to invest in sustainable tourism initiatives, hindering their capacity for development and expansion. Implementing environmentally sustainable practices requires investments and efforts (Al-Roubaie & Sarea, 2019), and SHGs may encounter challenges in adopting eco-friendly measures due to resource constraints or lack of awareness (Asmoro & Resmiatini, 2021; Singhania et al., 2021). The financial constraints not only restrict the immediate implementation of eco-friendly measures but also pose a broader challenge to the overall developmental efforts of SHGs engaged in sustainable tourism. Addressing these financial constraints becomes imperative for unlocking the full potential of SHGs and fostering a more comprehensive and impactful contribution to sustainable tourism development.
- *Capacity and Skill Gaps:* SHGs face challenges in building the necessary capacities and acquiring the skills required for sustainable tourism practices (Nzama, 2010), including eco-friendly initiatives and responsible tourism management. Other challenges include insufficient knowledge about waste management, community involvement, entrepreneurial skills, language skills and management skills. Insufficient training and support contribute to the challenges faced by SHGs in effectively implementing sustainable tourism practices and meeting the expectations of tourists.
- *Market Access and Competition:* Community members encounter challenges related to limited access to markets and marketing opportunities (Asmoro & Resmiatini, 2021; Rangus et al., 2017). Despite their efforts in producing local

and indigenous products, SHGs members struggle to find markets to sell their goods, which affect their economic sustainability. Market linkage and access to tourism markets are crucial for SHGs to thrive and contribute effectively to sustainable tourism (Vinodan & Meera, 2022). Moreover, entry barriers and stiff competition in the tourism market pose challenges for SHGs in promoting and sustaining their initiatives, affecting their ability to attract visitors and generate revenue. Insufficient financial support and business knowledge limit their marketing and promotional efforts, making it difficult to compete against larger players in the tourism industry.

- *Regulatory Compliance:* Adhering to complex regulatory frameworks and meeting compliance standards for sustainable tourism practices becomes challenging for SHGs, especially if there is a lack of clarity or support from regulatory authorities. In order to address these challenges, it is crucial for SHGs to receive support from various stakeholders and government agencies. These challenges highlight the importance of providing capacity-building and training programs to SHGs (Chitale, 2017), specifically tailored to the needs of sustainable tourism. Regulatory compliance adds another layer of complexity. Understanding and adhering to sustainable tourism practices within the framework of intricate regulatory standards can be daunting for SHGs.
- *Lack of Awareness:* Ensuring active participation and support from local communities is crucial for sustainable tourism. However, SHGs face challenges in garnering community involvement, which is essential for the success of initiatives. Lack of awareness about tourism benefits and sustainable practices among community members hinders their engagement and cooperation with SHGs in implementing sustainable tourism initiatives. In addressing the challenges faced by SHGs in sustainable tourism, it is essential to emphasise the potential benefits that can be achieved through such initiatives. One of the key factors is the motivation to do fair business due to the deliberated social atmosphere. This highlights the importance of fostering a supportive environment that encourages SHGs to engage in sustainable tourism practices.
- *Infrastructure and Technology:* Inadequate infrastructure and limited access to technology may hinder the efficiency of SHG operations, impacting their ability to offer quality services and experiences to tourists (Vukovic et al., 2021). The inadequacies in essential infrastructure, such as transportation, communication and utility services, coupled with a lack of access to modern technologies, pose significant barriers for SHGs. These limitations hinder their capacity to effectively engage in tourism-related activities, obstructing the realisation of their full potential in fostering sustainable tourism development.
- *Readiness to Change:* SHGs often encounter resistance from traditional or established businesses and stakeholders when they attempt to engage in sustainable tourism initiatives. The willingness to change and adapt to new practices is crucial to overcome this resistance and foster positive relationships within the community. This resistance can be attributed to a variety of factors, such as prevailing business practices, entrenched interests and a reluctance to embrace new paradigms in the tourism industry. The readiness to change

among these established entities becomes a critical determinant in the successful integration of SHGs into sustainable tourism endeavours.

- *Sociocultural Constraints:* Deeply rooted sociocultural norms and practices within local communities can present challenges for SHGs in promoting sustainable tourism. For example, traditional gender roles, traditional hierarchies and societal expectations sometimes limit the participation of women in certain tourism-related activities or decision-making processes. Overcoming these barriers requires a comprehensive approach that addresses cultural sensitivities and promotes inclusivity within the SHGs.

8. Conclusion

The comprehensive analysis reveals the multifaceted contributions of SHGs to economic empowerment, cultural preservation, environmental conservation and social development, thereby significantly enhancing sustainable tourism at destinations. SHGs contribute to economic empowerment by creating job opportunities, generating income for local communities and fostering entrepreneurship. Additionally, they play a pivotal role in conserving cultural heritage, facilitating positive cultural exchanges between tourists and locals. Their noteworthy contributions extend to environmental conservation, mitigating adverse impacts on the natural environment. Furthermore, SHGs promote social development by fostering community cohesion, instilling self-confidence, enhancing decision-making abilities and providing training for skill development. By promoting sustainable tourism initiatives, SHGs exemplify a balanced approach that ensures the progress of local communities through tourism development without compromising the ecological and cultural integrity of the destination. Despite their substantial potential, SHGs encounter challenges such as financial constraints, skill gaps, limited market access and regulatory compliance issues. Factors like lack of awareness, insufficient infrastructure, technological limitations, resistance to change and sociocultural constraints further complicate their operations. Notwithstanding these challenges, SHGs present significant potential for community-driven sustainable tourism initiatives. Policymakers and stakeholders can unlock this potential by offering support, enacting favourable policies and addressing the challenges faced by these groups.

9. Future Research

Future research in this area should further explore recurring themes such as financial sustainability, capacity constraints and effective marketing strategies vital for the success of SHG-driven tourism initiatives. Comprehensive investigations into policy frameworks that support the seamless integration of SHGs into tourism planning and development are necessary to contribute to destination sustainability. Further research is also needed to propose practical solutions ensuring the long-term sustainability of SHG-driven tourism initiatives. Comparative studies across diverse cultural and geographical contexts would enrich the generalisability of findings. Moreover, longitudinal research could offer insights into the long-term sustainability and evolution of SHG-driven initiatives. Lastly, understanding the role of technology and digital platforms in enhancing the reach and effectiveness of SHGs in sustainable tourism could be a pertinent area for future investigation. The study's potential limitation lies in the reliance on secondary data, limiting the depth of firsthand insights and potentially missing context-specific nuances. Recognising and optimising the role of SHGs can substantially contribute to fostering a more sustainable, inclusive and community-centric tourism model.

References

Ahemad, S., & Katoch, O. R. (2022). Role of self-help groups (SHGs) in enhancing incomes of rural women in J&K, India. *South Asian Journal of Social Studies and Economics*, *16*(1), 24–32. https://doi.org/10.9734/sajsse/2022/v16i1602

Al-Roubaie, A., & Sarea, M. A. (2019). Green investment and sustainable development: The case of Islamic finance. *Journal of Islamic Business and Management (JIBM)*, *09*(01), 14–24. https://doi.org/10.26501/jibm/2019.0901-002

Artal-Tur, A., Villena-Navarro, M., & Alamá-Sabater, L. (2018). The relationship between cultural tourist behaviour and destination sustainability. *Anatolia*, *29*(2), 237–251. https://doi.org/10.1080/13032917.2017.1414444

Asmoro, B. T., & Resmiatini, E. (2021). Maximizing organizational structure of Sukodono tourism awareness group. *Proceedings of the 2nd Annual Conference on Social Science and Humanities (ANCOSH 2020)*, *542*(Ancosh 2020), 297–301. https://doi.org/10.2991/assehr.k.210413.068

Bwala, H. B., Oladosu, R. O., & Dukku, S. J. (2016). Role of self-help groups a needful approach to sustainable community development. *Journal of Culture, Society and Development*, *21*(2002), 94–98.

Carey, S., Gountas, Y., & Gilbert, D. (1997). Tour operators and destination sustainability. *Tourism Management*, *18*(7), 425–431. https://doi.org/10.1016/S0261-5177(97)00044-7

Chitale, C. C. M. (2017). A study of self help group at Tikona village. *Ramani Rajasekhar*, *8*(2), 154–159.

DHAN Foundation. (2023). *Tourism for development*. https://www.dhan.org/themes/dhan-tourism.php

Freitas, I. V. D., & Koskowski, M. (2021). Heritage and sustainable development: Capacity building through tourism. *Capacity Building Through Heritage Tourism*, 113–131. https://doi.org/10.1201/9781003034391-6

Gaas, A. O. (2019). Empowering the poor towards sustainable development—A case study of self-help group approach in Somaliland. *Open Journal of Social Sciences*, *07*(11), 26–37. https://doi.org/10.4236/jss.2019.711003

Ibnou-Laaroussi, S., Rjoub, H., & Wong, W. K. (2020). Sustainability of green tourism among international tourists and its influence on the achievement of green environment: Evidence from North Cyprus. *Sustainability*, *12*(14), 1–24. https://doi.org/10.3390/su12145698

Jayant, J., & Dev Bhatt, C. (2021). An overview of sustainable development through micro finance: A study of self-help group – Bank linkage program in central region of India. *Journal of University of Shanghai for Science and Technology*, *23*(05), 294–318. https://doi.org/10.51201/jusst/21/05139

Kalra, R. K., Anil, B., Tonts, M., & Siddique, K. H. M. (2013). Self-help groups in Indian agriculture: A case study of farmer groups in Punjab, Northern India. *Agroecology and Sustainable Food Systems*, *37*(5), 509–530. https://doi.org/10.1080/10440046.2012.719853

Kandpal, R., & Saizen, I. (2022). Self-help group participation towards sustainable solid waste management in peri-urban villages: Evidence from Mumbai Metropolitan Region, India. *Environment, Development and Sustainability*, *24*(3), 3791–3814. https://doi.org/10.1007/s10668-021-01588-6

Kumar, A., & Jasheena, C. J. (2016). Kudumbashree: Promoting the self-help group model of empowerment through women entrepreneurship in Kerala – A study. *Prabandhan: Indian Journal of Management*, *9*(6), 20–34. https://doi.org/10.17010/pijom/2016/v9i6/94959

Kumar, N., Raghunathan, K., Arrieta, A., Jilani, A., & Pandey, S. (2021). The power of the collective empowers women: Evidence from self-help groups in India. *World Development*, *146*, 105579. https://doi.org/10.1016/j.worlddev.2021.105579

Kumari, J. K. (2014). Woman empowerment through entrepreneurship in service sector with special reference to SHGs in tourism. *Global Journal for Research Analysis*, *III*(IX), 159–160. http://theglobaljournals.com/gra/file.php?val=September_2014_1410773806__58.pdf$##$http://theglobaljournals.com/gra/articles.php?val=MjM1NQ==&b1=229&k=58

Lee, T. H. (2013). Influence analysis of community resident support for sustainable tourism development. *Tourism Management*, *34*, 37–46. https://doi.org/10.1016/J.TOURMAN.2012.03.007

Lim, J. E., & Lee, H. R. (2020). Living as residents in a tourist destination: A phenomenological approach. *Sustainability*, *12*(5), 1–12. https://doi.org/10.3390/su12051836

Maina, P. (2021). Influence of women's participation in self-help groups on their income in Tigania West Sub-County, Kenya. *European Journal of Social Sciences Studies*, *6*(2), 64–76. https://doi.org/10.46827/ejsss.v6i2.1005

Mathew, P. V., & Sreejesh, S. (2017). Impact of responsible tourism on destination sustainability and quality of life of community in tourism destinations. *Journal of Hospitality and Tourism Management*, *31*, 83–89. https://doi.org/10.1016/J.JHTM.2016.10.001

Mishra, S., & Mishra, J. M. (2023). Self-help groups and community-based tourism: A study of interrelationship and convolutions. *Journal of Tourism*, *24*(1), 71.

Mongabay. (2020). A group of women protect Sindhudurg's mangroves through eco-tourism. https://india.mongabay.com/2020/11/a-group-of-women-protect-sindhudurgs-mangroves-through-eco-tourism/

Nadalipour, Z., Imani Khoshkhoo, M. H., & Eftekhari, A. R. (2019). An integrated model of destination sustainable competitiveness. *Competitiveness Review*, *29*(4), 314–335. https://doi.org/10.1108/CR-12-2017-0086

Nanda Praptiwi, E., Hilman, Y. A., Nasution, R. D., & Widiyahseno, B. (2022). Partnership model of the village authority in developing tourism sector in Conto Village, Bulukerto Sub-District, Wonogiri Regency. *Jurnal Ilmiah Dinamika Sosial*, *6*(2), 136–148. https://doi.org/10.38043/jids.v6i2.3581

Nichols, C. (2021). Self-help groups as platforms for development: The role of social capital. *World Development*, *146*, 105575. https://doi.org/10.1016/j.worlddev.2021.105575

Nzama, T. (2010). Challenges of sustainable rural tourism development in KwaZulu-Natal. *Inkanyiso: Journal of Humanities and Social Sciences, 2*(1), 44–53. https://doi.org/10.4314/ijhss.v2i1.62121

Pan, S. Y., Gao, M., Kim, H., Shah, K. J., Pei, S. L., & Chiang, P. C. (2018). Advances and challenges in sustainable tourism toward a green economy. *Science of the Total Environment, 635*, 452–469. https://doi.org/10.1016/j.scitotenv.2018.04.134

Rangus, M., Brumen, B., & Potočnik Topler, J. (2017). Sustainable tourism development in rural areas: The role of stakeholders. *Academica Turistica, 10*(2), 167–173. https://doi.org/10.26493/2335-4194.10.167-173

Ravindran, A. M., & Vinodan, A. (2009). Implication of microfinance on tourism development and inclusive growth in India: Emerging models. *South Asian Journal of Tourism & Heritage, 1*(2), 56–67.

Ruducha, J., Hariharan, D., Potter, J., Ahmad, D., Kumar, S., Mohanan, P. S., Irani, L., & Long, K. N. G. (2019). Measuring coordination between women's self-help groups and local health systems in rural India: A social network analysis. *BMJ Open, 9*(8), 1–13. https://doi.org/10.1136/bmjopen-2019-028943

Samal, A., & Mohopatra, A. K. D. (2016). The issues and challenges of self-help groups: A study in Odisha. *International Research Journal of Management Science and Technology, 7*(12), 79–86. https://www.academia.edu/34974744/THE_ISSUES_AND_CHALLENGES_OF_SELF_HELP_GROUPS_A_STUDY_IN_ODISHA?email_work_card=title

Samishetti, B., & Rao, T. U. M. (2018). Self-help groups (SHGS) role in empowerment of women in India. In *1st International Conference on Recent Trends in Engineering, Materials, Management and Sciences (ICRTEMMS-2018)*, SBIT, Khammam, India, 25–27 October 2018 (Vol. 978, pp. 625–629).

Simpson, K. (2001). Strategic planning and community involvement as contributors to sustainable tourism development. *Current Issues in Tourism*, February 2013, 37–41.

Singhania, O., Swain, S., & George, B. (2021). Women empowerment in a traditional masculine industry: Tourism in context. *Economics, Management and Sustainability, 6*(2), 60–71. https://doi.org/10.14254/jems.2021.6-2.5

Stefanica, M., & Vlavian-Gurmeza, M. (2010). Ecotourism – Model of sustainable tourist development. *Studies and Scientific Researches. Economics Edition, 15*, 15–18. https://doi.org/10.29358/sceco.v0i15.161

Than, T. T., Kieu, T. P. H., Pham, T. A. D., Van Hoang, T. C., Tran, T. H., Nguyen, H. D., & Dao, T. K. (2020). Impact of community attachment and resident's support on destination sustainability: Evidence from spiritual and community destination in Vietnam. *Journal of Asian Finance, Economics and Business, 7*(8), 361–369. https://doi.org/10.13106/JAFEB.2020.VOL7.NO8.361

Venugopalan, T. (2021). Tourism and sustainability in India – Exploring sustainability of Goa tourism from the perspective of local community. *European Journal of Business and Management Research, 6*(3), 34–41. https://doi.org/10.24018/ejbmr.2021.6.3.852

Vinodan, A., & Meera, S. (2022). Case study on self-help groups intervention and destination sustainability in Kerala: Input-output-outcome model. *Tourism, 70*(2), 201–222. https://doi.org/10.37741/T.70.2.4

Vukovic, D. B., Petrovic, M., Maiti, M., & Vujko, A. (2021). Tourism development, entrepreneurship and women's empowerment – Focus on Serbian countryside. *Journal of Tourism Futures*, 1–21. https://doi.org/10.1108/JTF-10-2020-0167

Wei, Q., Mimi, L., Honggen, X., & Jinhe, Z. (2021). Study on the influence of tourists' value on sustainable development of Huizhou traditional villages – A case of Hongcun and Xidi. *E3S Web of Conferences, 236*. https://doi.org/10.1051/e3sconf/202123603007

Printed and bound by CPI Group (UK) Ltd, Croydon, CR0 4YY

13/09/2026

14961063-0003